The Jessie and John Danz Lectures

THE JESSIE AND JOHN DANZ LECTURES

The Human Crisis, by Julian Huxley

Of Men and Galaxies, by Fred Hoyle

The Challenge of Science, by George Boas

Of Molecules and Men, by Francis Crick

Nothing But or Something More, by Jacquetta Hawkes

How Musical Is Man?, by John Blacking

Abortion in a Crowded World: The Problem of Abortion with Special Reference to India, by S. Chandrasekhar

World Culture and the Black Experience, by Ali Mazrui

Energy for Tomorrow, by Philip H. Abelson

Plato's Universe, by Gregory Vlastos

The Nature of Biography, by Robert Gittings

Darwinism and Human Affairs. by Richard D. Alexander

Arms Control and SALT II, by W. K. H. Panofsky

Promethean Ethics: Living with Death, Competition, and Triage, by Garrett Hardin

Social Environment and Health, by Stewart Wolf

The Possible and the Actual, by François Jacob

Facing the Threat of Nuclear Weapons, by Sidney D. Drell

Symmetries, Asymmetries, and the World of Particles, by T. D. Lee

The Symbolism of Habitat: An Interpretation of Landscape in the Arts, by Jay Appleton

The Essence of Chaos, by Edward N. Lorenz

Environmental Health Risks and Public Policy: Decision Making in Free Societies, by David V. Bates

THE ESSENCE OF CHAOS

Edward Lorenz

UNIVERSITY OF WASHINGTON PRESS
Seattle

First paperback edition, 1995
Printed in the United Kingdom
Published in the United Kingdom by UCL Press Limited,
University College London

Library of Congress Cataloging-in-Publication Data
Lorenz, Edward N.
The essence of chaos / Edward N. Lorenz
p. cm.—(The Jessie and John Danz lectures)
Includes bibliographical references and index.
ISBN 0-295-97514-8 (alk. paper)
1. Chaotic behavior in systems. I. Title II. Series.
Q172.5.C45L67 1993   93-1835
003'.7—dc20   CIP

The paper used in this publication meets the minimum requirements of American National Standard for Information Sciences—Permanence of Paper for Printed Library Materials, ANSI Z39.48–1984. ∞

The Jessie and John Danz Lectures

IN OCTOBER 1961, Mr. John Danz, a Seattle pioneer, and his wife, Jessie Danz, made a substantial gift to the University of Washington to establish a perpetual fund to provide income to be used to bring to the University of Washington each year "distinguished scholars of national and international reputation who have concerned themselves with the impact of science and philosophy on man's perception of a rational universe." The fund established by Mr. and Mrs. Danz is now known as the Jessie and John Danz Fund, and the scholars brought to the University under its provisions are known as Jessie and John Danz Lecturers or Professors.

Mr. Danz wisely left to the Board of Regents of the University of Washington the identification of the special fields in science, philosophy, and other disciplines in which lectureships may be established. His major concern and interest were that the fund would enable the University of Washington to bring to the campus some of the truly great scholars and thinkers of the world.

Mr. Danz authorized the Regents to expend a portion of the income from the fund to purchase special collections of books, documents, and other scholarly materials needed to reinforce the effectiveness of the extraordinary lectureships and professorships. The terms of the gift also provided for the publication and dissemination, when this seems appropriate, of the lectures given by the Jessie and John Danz Lecturers.

Through this book, therefore, another Jessie and John Danz Lecturer speaks to the people and scholars of the world, as he has spoken to his audiences at the University of Washington and in the Pacific Northwest community.

To my grandchildren, Nicky and Sarah

Contents

Preface

IN THE SPRING OF 1990 I received an invitation from the University of Washington to deliver a set of lectures, as part of the series that had been inaugurated a generation earlier through the benevolence and farsightedness of Jessie and John Danz. The lectures were to be delivered before a general audience, and I was free to choose a subject.

Some thirty years previously, while conducting an extensive experiment in the theory of weather forecasting, I had come across a phenomenon that later came to be called "chaos"—seemingly random and unpredictable behavior that nevertheless proceeds according to precise and often easily expressed rules. Earlier investigators had occasionally encountered behavior of this sort, but usually under rather different circumstances. Often they failed to recognize what they had seen, and simply became aware that something was blocking them from solving their equations or otherwise completing their studies. My situation was unique in that, as I eventually came to realize, my experiment was doomed to failure *unless* I could construct a system of equations whose solutions behaved chaotically. Chaos suddenly became something to be welcomed, at least under some conditions, and in the ensuing years I found myself turning more and more toward chaos as a phenomenon worthy of study for its own sake.

It was easy to decide what topic the lectures should cover. I accepted the invitation, and chose as a title "The Essence of Chaos." Eventually a set of three lectures took shape. The first one defined chaos and illustrated its basic properties with some simple examples, and ended by describing some related phenomena—nonlinearity, complexity, and fractality—that had also come to be called "chaos." The second lecture dealt with the global weather as a complicated example of a chaotic system. The final one presented an account of our growing awareness of chaos, offered a prescription via which one could design one's own chaotic systems, and ended with some philosophical speculations. In keeping with the anticipated make-up of the audience, I displayed no mathematical formulas, and avoided technical terms except for some that I defined as I went along.

The present volume, with the same title, is written in the spirit of the Danz Lectures. It contains the same material, together with additions written to fill in the many gaps that were inevitably present in a limited oral presentation. The leading lecture has been expanded to become Chapters 1, 2, and 5, while the second one has been made into Chapter 3. The final lecture, with its historical account that begins with the discovery of Neptune, proceeds through the work of Henri Poincaré and his successors, and pauses to tell of my own involvement with chaos, has become Chapter 4.

My decision to convert the lectures into a book has been influenced by my conviction that chaos, along with its many associated concepts—strange attractors, basin boundaries, period-doubling bifurcations, and the like—can readily be understood and relished by readers who have no special mathematical or other scientific background, despite the occasionally encountered references to chaos as a branch of mathematics or a new science. As in the lectures, I have presented the chaos story in nontechnical language, except where, to avoid excessive repetition of lengthy phrases, I have introduced and defined a number of standard terms. I have placed the relevant mathematical equations and their derivations in an appendix, which need not be read for an understanding of the main text, but which may increase the volume's appeal to the mathematically minded reader.

Of course one cannot maintain that there is no mathematics at all in the main text, except by adopting a rather narrow view of what constitutes mathematics. For example, merely noting that one illustration shows two boards sliding thirty meters down a slope, starting ten centimeters apart and ending up ten meters apart, can be looked upon as a mathematical observation; a verbal description of what the illustration depicts is then a mathematical statement. In any event, a good deal of less simple mathematics has gone into the production of the illustrations; most of them are end products of mathematical developments, subsequently converted into computer programs. The reader nevertheless need not confront the formulas, nor the programs, to be able to absorb the messages that the illustrations contain.

For their aid during the preparation of this work I am indebted to many persons. First of all I must thank the Danz Foundation, without whose sponsorship of my lectures I would never have taken the first step. I must likewise thank the University of Washington for choosing me as a lecturer. I am deeply indebted to the Climate Dynamics Program of the Atmospheric Sciences Section of the National Science Foundation, and to the program's current director, Jay Fein, for supporting my research in chaos and its applications to the atmosphere over many years, and, most immediately, for making it possible for me to write the numerous computer programs and to perform the subsequent computations that have resulted in the illustrations in this volume. I wish to thank Joel Sloman for typing and otherwise assisting with not only the final manuscript but also the innumerable intermediate versions, Diana Spiegel for her ever-present aid in dealing with the vagaries of our computer system, and Jane McNabb for bearing the bulk of the administrative burden that otherwise would have fallen on me.

Thanks go to Dave Fultz of the University of Chicago for supplying the photographs of his dishpan experiments, and to him and the American Meteorological Society for permission to reproduce them. Thanks also go to Robert Dattore and Wilbur Spangler of the National Center for Atmospheric Research for preparing and making available the lengthy tape containing the many years of recorded upper-level weather data at Singapore.

I must give special recognition to Merry Caston, who has gone over the manuscript page by page, and whose pertinent comments have led me to incorporate a good many clarifying additions and other amendments. There are many other persons with whom I have had brief or in some cases extensive conversations, which have exerted their influence on the words that I have written or the ideas that I have expressed. In this connection I must particularly mention Robert Cornett, James Curry, Robert Devaney, Alan Faller, Robert Hilborn, Philip Merilees, Tim Palmer, Bruce Street, Yoshisuke Ueda, J. Michael Wallace, and James Yorke. To still others who may have similarly influenced me without my being aware of it, and also to some anonymous reviewers, I can only say that their names ought to have been included.

Finally, I am most grateful to my wife, Jane, who has supplied moral support throughout the preparation of this volume and has accompanied me on numerous travels in search of chaotic material, and to my children Nancy, Edward, and Cheryl—lawyer, economist, and psychologist—who have perfectly filled the role of the intelligent layperson and have subjected the manuscript in various stages of completion to their closest scrutiny.

THE ESSENCE OF CHAOS

CHAPTER 1

Glimpses of Chaos

It Only Looks Random

WORDS are not living creatures; they cannot breathe, nor walk, nor become fond of one another. Yet, like the human beings whom they are destined to serve, they can lead unique lives. A word may be born into a language with just one meaning, but, as it grows up, it may acquire new meanings that are related but nevertheless distinct. Often these meanings are rather natural extensions of older ones. Early in our own lives we learn what "hot" and "cold" mean, but as we mature we discover that hot pursuit and cold comfort, or hot denials and cold receptions, are not substances or objects whose temperatures can be measured or estimated. In other instances the more recent meanings are specializations. We learn at an equally early age what "drink" means, but if later in life someone says to us, "You've been drinking," we know that he is not suggesting that we have just downed a glass of orange juice. Indeed, if he tells someone else that we drink, he is probably implying not simply that we often consume alcoholic beverages, but that we drink enough to affect our health or behavior.

So it is with "chaos"—an ancient word originally denoting a complete lack of form or systematic arrangement, but now often used to imply the absence of some kind of order that ought to be present. Notwithstanding its age, this familiar word is not close to its deathbed, and it has recently outdone many other common words by acquiring several related but distinct *technical* meanings.

It is not surprising that, over the years, the term has often been used by various scientists to denote randomness of one sort or another. A recent example is provided by the penetrating book *Order Out of Chaos*, written by the Nobel Prize–winning physical chemist Ilya Prigogine and his colleague Isabelle Stengers. These authors deal with the manner in which many disorganized systems can spontaneously acquire organization, just as a shapeless liquid mass can, upon cooling, solidify into an exquisite crystal. A generation or two earlier, the mathematician Norbert Wiener would sometimes even pluralize the word, and would write about a chaos or several chaoses when referring to systems like the host of randomly located molecules that form a gas, or the haphazardly arranged collection of water droplets that make up a cloud.

This usage persists, but, since the middle 1970s, the term has also appeared more and more frequently in the scientific literature in one or another of its recently acquired senses; one might well say that there are several newly named kinds of chaos. In this volume we shall be looking closely at one of them. There are numerous processes, such as the swinging of a pendulum in a clock, the tumbling of a rock down a mountainside, or the breaking of waves on an ocean shore, in which variations of some sort take place as time advances. Among these processes are some, perhaps including the rock and the waves but omitting the pendulum, whose variations are *not random but look random.* I shall use the term *chaos* to refer collectively to processes of this sort—ones that appear to proceed according to chance even though their behavior is in fact determined by precise laws. This usage is arguably the one most often encountered in technical works today, and scientists writing about chaos in this sense no longer feel the need to say so explicitly.

In reading present-day accounts, we must keep in mind that one of the other new usages may be intended. Sometimes the phenomena being described are things that appear to have random arrangements in space rather than random progressions in time, like wildflowers dotting a field. On other occasions, the arrangements or progressions are simply very intricate rather than seemingly random, like the pattern woven into an oriental rug. The situation is further complicated because several other terms, notably *nonlinearity, complexity,* and *fractality,* are often used more or less synonymously with *chaos* in one or several of its senses. In a later chapter I shall have a bit to say about these related expressions.

In his best-selling book *Chaos: Making a New Science,* which deals with chaos in several of its newer senses, James Gleick suggests that chaos theory may in time rival relativity and quantum mechanics in its influence on scientific thought. Whether or not such a prophecy comes true, the "new science" has without question jumped into the race with certain advantages. Systems that presumably qualify as examples of chaos can very often be seen and appreciated without telescopes or microscopes, and they can be recorded without time-lapse or high-speed cameras. Phenomena that are supposedly chaotic include simple everyday occurrences, like the falling of a leaf or the flapping of a flag, as well as much more involved processes, like the fluctuations of climate or even the course of life itself.

I have said "presumably" and "supposedly" because there is something about these phenomena that is not quite compatible with my description of chaos as something that is random in appearance only. Tangible physical systems generally possess at least a small amount of true randomness. Even the seemingly regular swinging of the pendulum in a cuckoo clock may in reality be slightly disturbed by currents in the air or vibrations in the wall; these may in turn be produced by people moving about in a room or traffic passing down a nearby street. If chaos consists of things that are actually *not* random and only *seem* to be, must it exclude familiar everyday phenomena that have a bit of randomness, and be confined to mathematical abstractions? Might not such a restriction severely diminish its universal significance?

An acceptable way to render the restriction unnecessary would be to stretch the definition of chaos to include phenomena that are slightly random, provided that their much greater apparent randomness is not a by-product of their slight true randomness. That is, real-world processes that appear to be behaving randomly—perhaps the falling leaf or the flapping flag—should be allowed to qualify as chaos, as long as they would continue to appear random even if any true randomness could somehow be eliminated.

In practice, it may be impossible to purge a real system of its actual randomness and observe the consequences, but often we can guess what these would be by turning to theory. Most theoretical studies of real phenomena are studies of approximations. A scientist attempting to explain the motion of a simple swinging pendulum, which incidentally is not a chaotic system, is likely to neglect any extraneous random vibrations and air currents, leaving such considerations to the more practical engineer. Often he or she will even disregard the clockwork that keeps the pendulum swinging, and the internal friction that makes the clockwork necessary, along with anything else that is inconvenient. The resulting pencil-and-paper system will be only a model, but one that is completely manageable. It seems appropriate to call a real physical system chaotic if a fairly realistic model, but one with the system's inherent randomness suppressed, still *appears* to behave randomly.

Pinballs and Butterflies

My somewhat colloquial definition may capture the essence of chaos, but it would cause many mathematicians to shudder. Probably most people in other walks of life are unaware of the extent to which mathematics is dependent on definitions. Whether or not a proposition as stated is true often depends upon just how the words contained in it have been defined. Certainly, before one can develop a rigorous theory dealing with some phenomenon, one needs an unambiguous definition of the phenomenon.

In the present instance the colloquial definition is ambiguous because "randomness" itself has two rather different definitions, although, as we shall presently see, this flaw can easily be removed by specifying the one that is intended. More serious is the simple expression "looks random," which does not belong in a rigorous discussion, since things that look alike to one person often do not look alike to another. Let us try to arrive at a working definition of chaos while retaining the spirit of the colloquial one.

According to the narrower definition of randomness, a *random* sequence of events is one in which anything that can ever happen can happen next. Usually it is also understood that the probability that a given event will happen next is the same as the probability that a like event will happen at any later time. A familiar example, often serving as a paradigm for randomness, is the toss of a coin. Here either heads or tails, the only two things that can ever happen, can happen next. If the process is indeed random, the probability of throwing heads on the next toss of any particular coin, whether 50 percent or something else, is precisely the same as that of throwing heads on any other toss of the same coin, and it will remain the same unless we toss the coin so violently that it is bent or worn out of shape. If we already know the probability, knowing in addition the outcome of the last toss cannot improve our chances of guessing the outcome of the next one correctly.

It is true that knowing the results of enough tosses of the same coin can suggest to us what the probability of heads is, for that coin, if we do not know it already. If after many tosses of the coin we become aware that heads has come up 55 percent of the time, we may suspect that the coin is biased, and that the probability has been, is, and will be 55 percent, rather than the 50 percent that we might have presupposed.

The coin is an example of complete randomness. It is the sort of randomness that one commonly has in mind when thinking of random numbers, or deciding to use a random-number generator. According to the broader definition of randomness, a *random* sequence is simply one in which any one of several things can happen next, even though not necessarily *anything* that can ever happen can happen next. What actually is possible next will then depend upon what has just happened. An example, which, like tossing a coin, is intimately associated with games of chance, is the shuffling of a deck of cards. The process is presumably random, because even if the shuffler should wish otherwise—for example, if on each riffle he planned to cut the deck exactly in the middle, and then allow a single card to fall on the table from one pile, followed by a single card from the other pile, etc.—he probably could not control the muscles in his fingers with sufficient precision to do so, unless he happened to be a virtuoso shuffler from a gaming establishment. Yet the process is not completely random, if by what happens next we mean the outcome of the next single riffle, since one riffle cannot change any given order of the cards in the deck to any other given order. In particular, a single riffle cannot completely reverse the order of the cards, although a sufficient number of successive riffles, of course, can produce any order.

A *deterministic* sequence is one in which only one thing can happen next; that is, its evolution is governed by precise laws. Randomness in the broader sense is therefore identical with the absence of determinism. It is this sort of randomness that I have intended in my description of chaos as something that *looks* random.

Tossing a coin and shuffling a deck are processes that take place in discrete steps—successive tosses or riffles. For quantities that vary continuously, such as the speed of a car on a highway, the concept of a next event appears to lose its meaning. Nevertheless, one can still define randomness in the broader sense, and say that it is present when more than one thing, such as more than one prespecified speed of a car, is possible at any specified future time. Here we may anticipate that the closer the future time is to the present, the narrower the range of possibilities—a car momentarily stopped in heavy traffic may be exceeding the speed limit ten seconds later, but not one second later. Mathematicians have found it advantageous to introduce the concept of a *completely* random continuous process, but it is hard to picture what such a process in nature might look like.

Systems that vary deterministically as time progresses, such as mathematical models of the swinging pendulum, the rolling rock, and the breaking wave, and also systems that vary with an inconsequential amount of randomness—possibly a real pendulum, rock, or wave—are technically known as *dynamical systems.* At least in the case of the models, the *state* of the system may be specified by the numerical values of one or more *variables.* For the model pendulum, two variables—the position and speed of the bob—will suffice; the speed is to be considered positive or negative, according to the direction in which the bob is currently moving. For the model rock, the position and velocity are again required, but, if the model is to be more realistic, additional variables that specify the orientation and spin are needed. A breaking wave is so intricate that a fairly realistic model would have to possess dozens, or more likely hundreds, of variables.

Returning to chaos, we may describe it as behavior that *is* deterministic, or is nearly so if it occurs in a tangible system that possesses a slight amount of randomness, but does not *look* deterministic. This means that the present state completely or almost completely determines the future, but does not appear to do so. How can deterministic behavior look random? If truly identical states do occur on two or more occasions, it is unlikely that the identical states that will necessarily follow will be perceived as being appreciably different. What can readily happen instead is that almost, but not quite, identical states occurring on two occasions will *appear* to be just alike, while the states that follow, which need not be even nearly alike, will be observably different. In fact, in some dynamical systems it is normal for two almost identical states to be followed, after a sufficient time lapse, by two states bearing no more resemblance than two states chosen at random from a long sequence. Systems in which this is the case are said to be *sensitively dependent on initial conditions.* With a few more qualifications, to be considered presently, sensitive dependence can serve as an acceptable definition of chaos, and it is the one that I shall choose.

"Initial conditions" need not be the ones that existed when a system was created. Often they are the conditions at the beginning of an experiment or a computation, but they may also be the ones at the beginning of any stretch of time that interests an investigator, so that one person's initial conditions may be another's midstream or final conditions.

Sensitive dependence implies more than a mere increase in the difference between two states as each evolves with time. For example, there are deterministic systems in which an initial difference of one unit between two states will eventually increase to a hundred units, while an initial difference of a hundredth of a unit, or even a millionth of a unit, will eventually increase to a hundred units also, even though the latter increase will inevitably consume more time. There are other deterministic systems in which an initial difference of one unit will increase to a hundred units, but an initial difference of a hundredth of a unit will increase only to one unit. Systems of the former sort are regarded as chaotic, while those of the latter sort are not considered to constitute chaos, even though they share some of its properties.

Because chaos is deterministic, or nearly so, games of chance should not be expected to provide us with simple examples, but games that *appear* to involve chance ought to be able to take their place. Among the devices that can produce chaos, the one that is nearest of kin to the coin or the deck of cards may well be the pinball machine. It should be an old-fashioned one, with no flippers or flashing lights, and with nothing but simple pins to disturb the free roll of the ball until it scores or becomes dead.

One spring in the thirties, during my undergraduate years at Dartmouth, a few pinball machines suddenly appeared in the local drugstores and eating places. Soon many students were occasionally winning, but more often losing, considerable numbers of nickels. Before long the town authorities decided that the machines violated the gambling laws and would have to be removed, but they were eventually persuaded, temporarily at least, that the machines were contests of skill rather than games of chance, and were therefore perfectly legal.

If this was indeed so, why didn't the students perfect their skill and become regular winners? The reason was chaos. As counterparts of successive tosses of a coin or riffles of a deck, let the "events" be successive strikes on a pin. Let the outcome of an event consist of the particular pin that is struck, together with the direction from the pin to the center of the ball, and the velocity of the ball as it leaves the pin. Note that I am using *velocity* in the technical sense, to denote speed together with direction of motion, just as *position* with respect to some reference point implies distance together with direction of displacement.

Suppose that two balls depart one after the other from the same pin in slightly different directions. When the balls arrive at the next pin, their positions will be close together, compared to the distance between the pins, but not necessarily close, compared to the diameter of a ball. Thus, if one ball hits the pin squarely and rebounds in the direction from which it came, the other can strike it obliquely and rebound at right angles. This is approximately what happens in Figure 1, which shows the paths of the centers of two balls that have left the plunger of a pinball machine at nearly equal speeds. We see that the angle between two paths can easily increase tenfold whenever a pin is struck, until soon one ball will completely miss a pin that the other one hits. Thus a player will need to increase his or her control tenfold in order to strike one more pin along an intended pathway.

Of course, the pinball machine in Figure 1 is really a mathematical model, and the paths of the balls have been computed. The model has incorporated the decelerating effect of friction, along with a further loss of energy whenever a ball bounces from a pin or a side wall, but, in a real machine, a ball will generally acquire some side spin as it hits a pin, and this will alter the manner in which it will rebound from the next pin. It should not alter the conclusion that the behavior is chaotic—that the path is sensitively dependent on the initial speed.

Even so, the model as it stands fails in one respect to provide a perfect example of chaos, since the chaotic behavior ceases after the last pin is struck. If, for example, a particular ball hits only seven pins on its downward journey, a change of a millionth of a degree in its initial direction would amplify to ten degrees, but a change of a ten-millionth of a degree would reach only one degree. To satisfy all of the requirements for chaos, the machine would have to be infinitely long—a possibility in a model if not in reality—or else there would have to be some other means of keeping the ball in play forever. Any change in direction, even a millionth of a millionth of a degree, would then have the opportunity to amplify beyond ten degrees.

An immediate consequence of sensitive dependence in any system is the impossibility of making perfect predictions, or even mediocre pre-

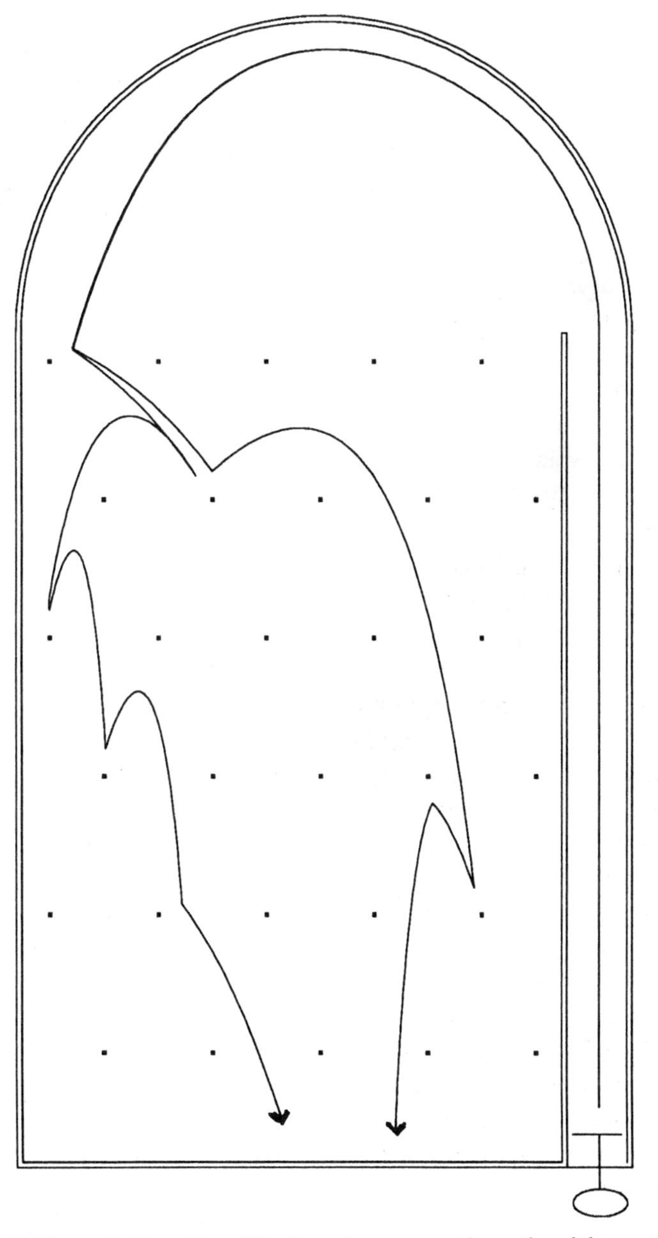

Figure 1. The pinball machine. The jagged curves are the paths of the centers of two balls that have begun their journeys at nearly equal speeds. The radius of a ball is indicated by the distance between a pin and an abrupt change in the direction of a path.

dictions sufficiently far into the future. This assertion presupposes that we cannot make measurements that are completely free of uncertainty. We cannot estimate by eye, to the nearest tenth of a degree nor probably to the nearest degree, the direction in which a pinball is moving. This means that we cannot predict, to the nearest ten degrees, the ball's direction after one or two strikes on a pin, so that we cannot even predict which pin will be the third or fourth to be struck. Sophisticated electronic equipment might measure the direction to the nearest thousandth of a degree, but this would merely increase the range of predictability by two or three pins. As we shall see in a later chapter, sensitive dependence is also the chief cause of our well-known failure to make nearly perfect weather forecasts.

I have mentioned two types of processes—those that advance step by step, like the arrangements of cards in a deck, and those that vary continuously, like the positions or speeds of cars on a highway. As dynamical systems, these types are by no means unrelated. The pinball game can serve to illustrate a fundamental connection between them.

Suppose that we observe 300 balls as they travel one by one through the machine. Let us construct a diagram containing 300 points. Each point will indicate the position of the center of one ball when that ball strikes its first pin. Let us subsequently construct a similar diagram for the second strike. The latter diagram may then be treated as a full-scale map of the former, although certainly a rather distorted map. A very closely spaced cluster of points in the first diagram may appear as a recognizable cluster in the second. Dynamical systems that vary in discrete steps, like the pinball machine whose "events" are strikes on a pin, are technically known as *mappings*. The mathematical tool for handling a mapping is the *difference equation*. A system of difference equations amounts to a set of formulas that together express the values of all of the variables at the next step in terms of the values at the current step.

I have been treating the pinball game as a sequence of events, but of course the motion of a ball between strikes is as precisely governed by physical laws as are the rebounds when the strikes occur. So, for that matter, is the motion of a coin while it is in the air. Why should the latter process be randomness, while the former one is chaos? Between any two coin tosses there is human intervention, so that the outcome of one toss fails to determine the outcome of the next. As for the ball, the only human influence on its path occurs before the first pin is struck, unless the player has mastered the art of jiggling the machine without activating the tilt sign.

Since we can observe a ball between strikes, we have the option of plotting diagrams that show the positions of the centers of the 300 balls at a sequence of closely spaced times, say every fiftieth of a second, instead of only at moments of impact. Again each diagram will be a full-scale map of the preceding one. Now, however, the prominent features will be only slightly changed from one diagram to the next, and will appear to flow through the sequence. Dynamical systems that vary continuously, like the pendulum and the rolling rock, and evidently the pinball machine when a ball's complete motion is considered, are technically known as *flows*. The mathematical tool for handling a flow is the *differential equation*. A system of differential equations amounts to a set of formulas that together express the rates at which all of the variables are currently changing, in terms of the current values of the variables.

When the pinball game is treated as a flow instead of a mapping, and a simple enough system of differential equations is used as a model, it may be possible to solve the equations. A complete solution will contain expressions that give the values of the variables at any given time in terms of the values at any previous time. When the times are those of consecutive strikes on a pin, the expressions will amount to nothing more than a system of difference equations, which in this case will have been derived by solving the differential equations. Thus a mapping will have been derived from a flow.

Indeed, we can create a mapping from *any* flow simply by observing the flow only at selected times. If there are no special events, like strikes on a pin, we can select the times as we wish—for instance, every hour on the hour. Very often, when the flow is defined by a set of differential equations, we lack a suitable means for solving them—some differential equations are intrinsically unsolvable. In this event, even though the difference equations of the associated mapping must exist as relationships, we cannot find out what they look like. For some real-world systems we even lack the knowledge needed to formulate the differential equations; can we honestly expect to write any equations that realistically describe surging waves, with all their bubbles and spray, being driven by a gusty wind against a rocky shore?

If the pinball game is to chaos what the coin toss is to complete randomness, it has certainly not gained the popularity as a symbol for chaos

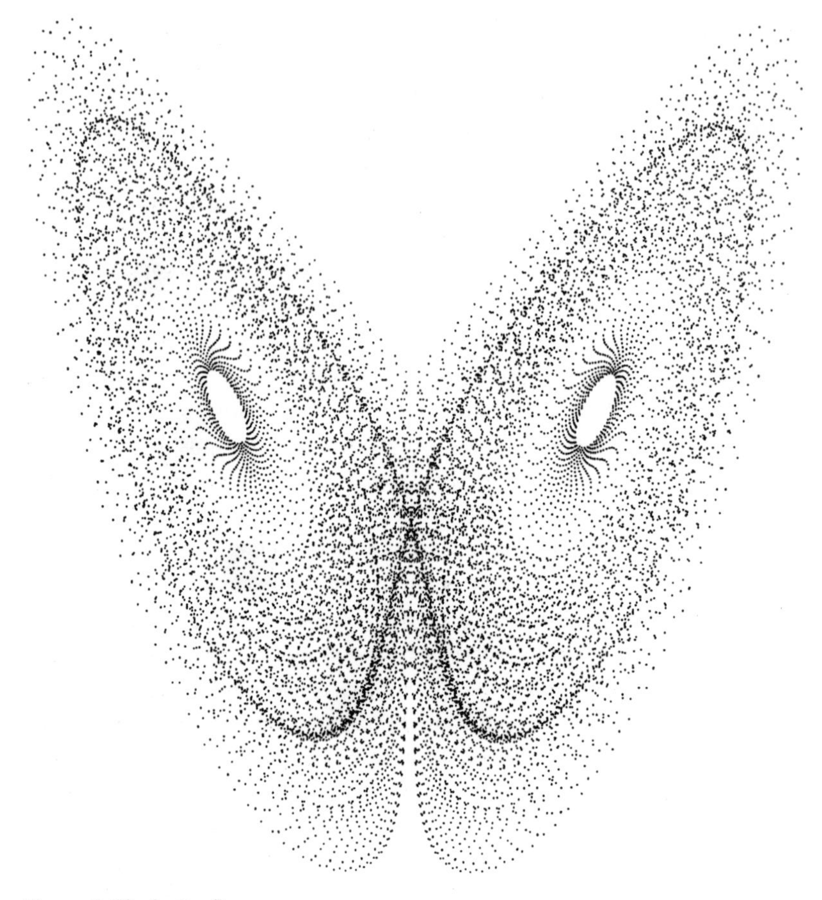

Figure 2. The butterfly.

that the coin has enjoyed as a symbol for randomness. That distinction at present seems to be going to the butterfly, which has easily outdistanced any potential competitors since the appearance of James Gleick's book, whose leading chapter is entitled "The Butterfly Effect."

The expression has a somewhat cloudy history. It appears to have arisen following a paper that I presented at a meeting in Washington in 1972, entitled "Does the Flap of a Butterfly's Wings in Brazil Set Off a Tornado in Texas?" I avoided answering the question, but noted that if a single flap could lead to a tornado that would not otherwise have formed, it could equally well prevent a tornado that would otherwise have formed. I noted also that a single flap would have no more effect on the weather than any flap of any other butterfly's wings, not to mention the activities of other species, including our own. The paper is reproduced in its original form as Appendix 1.

The thing that has made the origin of the phrase a bit uncertain is a peculiarity of the first chaotic system that I studied in detail. Here an abbreviated graphical representation of a special collection of states known as a "strange attractor" was subsequently found to resemble a butterfly, and soon became known as the butterfly. In Figure 2 we see one butterfly; a representative of a closely related species appears on the inside cover of Gleick's book. A number of people with whom I have talked have assumed that the butterfly effect was named after this attractor. Perhaps it was.

Some correspondents have also called my attention to Ray Bradbury's intriguing short story "A Sound of Thunder," written long before the Washington meeting. Here the death of a prehistoric butterfly, and its consequent failure to reproduce, change the outcome of a present-day presidential election.

Before the Washington meeting I had sometimes used a sea gull as a symbol for sensitive dependence. The switch to a butterfly was actually made by the session convenor, the meteorologist Philip Merilees, who was unable to check with me when he had to submit the program titles. Phil has recently assured me that he was not familiar with Bradbury's story. Perhaps the butterfly, with its seeming frailty and lack of power, is a natural choice for a symbol of the small that can produce the great.

Other symbols have preceded the sea gull. In George R. Stewart's novel *Storm,* a copy of which my sister gave me for Christmas when she first learned that I was to become a meteorology student, a meteorologist recalls his professor's remark that a man sneezing in China may set people to shoveling snow in New York. Stewart's professor was simply echoing what some real-world meteorologists had been saying for many years, sometimes facetiously, sometimes seriously.

It Ain't Got Rhythm

There are a number of heart conditions known as arrhythmias; some of them can be fatal. The heart will proceed to beat at irregular intervals and sometimes with varying intensity, instead of ticking like a metronome. It has been conjectured that arrhythmias are manifestations of chaos. Clearly they entail an absence of some order that ought to be present, but let us see how they, or any processes that lack rhythm, may be related to sensitive dependence.

Precise definitions are not always convenient ones. Having defined chaos in terms of sensitive dependence, we may discover that it is difficult to determine whether certain phenomena are chaotic.

The pinball machine should present no problem. If we have watched a ball as it rolls, and have noted its position and velocity at some "initial" time, it should be fairly easy for us to set a new ball rolling from about the same position with about the same speed and direction, and then see whether it follows almost the same path; presumably it will not, if my analysis has been correct. If instead our system is a flag flapping in the wind, we do not have this option. We might record the flag's behavior for a while by taking high-speed photographs, but it would be difficult to bend the flag into the shape appearing on a selected exposure, especially with a good wind blowing, and even harder to set each point of the flag moving with the appropriate velocity.

Perhaps on two occasions we might hold the flag taut with some stout string, and let the initial moments be the times when the strings are cut. If we then worry that subsequent differences in behavior may result from fluctuations in the wind instead of from intrinsic properties of the flag, we can circumvent the problem by substituting an electric fan. Nevertheless, we shall have introduced highly unusual initial states—flags in a breeze do not ordinarily become taut—and these states may proceed to evolve in a highly unusual manner, thereby invalidating our experiment as a test for chaos.

Fortunately there is a simpler approach to systems like the flapping flag, and it involves looking for rhythm. Before we can justify it, we must examine a special property of certain dynamical systems, which is known as *compactness.*

Suppose that on a round of golf you reach the tee of your favorite par-three hole and drive your ball onto the green. If you should decide to drive a second ball, can you make it come to rest within a foot of the first one, if this is what you wish? Presumably not; even without any wind the needed muscular control is too great, and the balls might not be equally resilient. If instead you have several buckets of balls and continue to drive, you will eventually place a ball within a foot of one that you have already driven, although perhaps not close to the first one.

This will not be because your game will have improved during the day, but simply because the green, or even the green plus a few sand traps and a pond, cannot hold an unlimited number of balls, each one of which is at least a foot from any other. The argument still holds if the critical distance between the centers of two balls is one centimeter, or something still smaller, as long as your caddy removes each ball and marks the spot before the next drive. Of course you may need many more buckets of balls.

The surface of the green is two-dimensional—a point on it may be specified by its distance and direction from the cup—and its area is bounded. Many dynamical systems are like the ball on the green, in that their states may be specified by the values of a finite number of quantities, each of which varies within strict bounds. If in observing one of these systems we wait long enough, we shall eventually see a state that nearly duplicates an earlier one, simply because the number of possible states, no one of which closely resembles any other one, is limited. Systems in which arbitrarily close repetitions—closer than any prespecified degree—must eventually occur are called *compact*.

For practical purposes the flag is a compact system. The bends in the flag as it flaps often resemble smooth waves. Let us define the state of the flag by the position and velocity of each point of a well-chosen grid, perhaps including the centers of the stars if it is an American flag, instead of using every point on the flag. Two states that, by this definition, are nearly alike must then eventually occur, and any reasonable interpolation will indicate that the positions and velocities of any other points on the flag will also be nearly alike in the two instances.

Our pinball machine is not a compact system. Not only do near repetitions not have to occur while a single ball is in play, but they cannot, since friction is continually tending to slow the ball, and the only way that the ball can regain or maintain its speed is to roll closer to the base of the machine. However, we can easily visualize a modified system where near repetitions are inevitable.

Imagine a very long pinball machine; it might stand on a gently sloping sidewalk outside a local drugstore, and extend for a city block. Let the playing surface be marked off into sections, say one meter long, and let the arrangement of the pins in each section be identical with that in any other. Except for being displaced from each other by one or more sections, the complete paths of two balls, occupying similar positions in different sections, and moving with identical velocities—speeds and directions—will be identical also. Thus for practical purposes the state of the system may be defined by the ball's velocity, together with its position with respect to a key point in its section, such as the uppermost pin. These quantities all vary within limited ranges. It follows that if the city block is long enough, a near repetition of a previous state must eventually occur.

In Figure 3 we see the computed path of the center of a single ball as it works its way past eighty pins, in a machine that is not only elongated but very narrow; the playing space is only twice as wide as the ball. Each pin is set one-quarter of the way in from one side wall or the other. The long machine is displayed as four columns; the right-hand column contains the first twenty pins, and each remaining column is an extension of the one to its right. The vertical scale has been compressed, as if we were looking from a distance with our eye just above the playing surface, so that the circular upper end appears elliptical, as would an area directly below the ball, if it were shown. You may find the figure easier to study if you give it a quarter turn counterclockwise and look at the path as a graph. Evidently the ball completely misses about a third of the pins, but the rebounds from pins somewhat outnumber those from side walls.

We see that the path from above the first to below the seventh pin in the third column nearly duplicates the path in the same part of the second column. With the initial speed used in the computation, the expected close repetition of an earlier state has set in only forty pins from the start. If the width of the playing space had not been so drastically restricted, thus limiting the possible positions of the ball, we could have expected a much longer wait, but not an eternal one.

Suppose now that some compact system, perhaps an elongated pinball machine or perhaps a flag, is *not* chaotic; that is, it does not exhibit sensitive dependence. Assume that the inevitable very near repetition of an "initial" state occurs after ten seconds, although it could be after an hour or longer. From that time onward, the behavior of the system will nearly repeat the behavior that occurred ten seconds earlier, until, after ten more seconds, another near repetition of the initial state will occur. After another ten seconds there will be still another near repetition, and so on, and ultimately the behavior may become *periodic;* that is, a particular pattern of behavior may be repeated over and over again, in this case at ten-second intervals. It is also possible that a true ten-second pe-

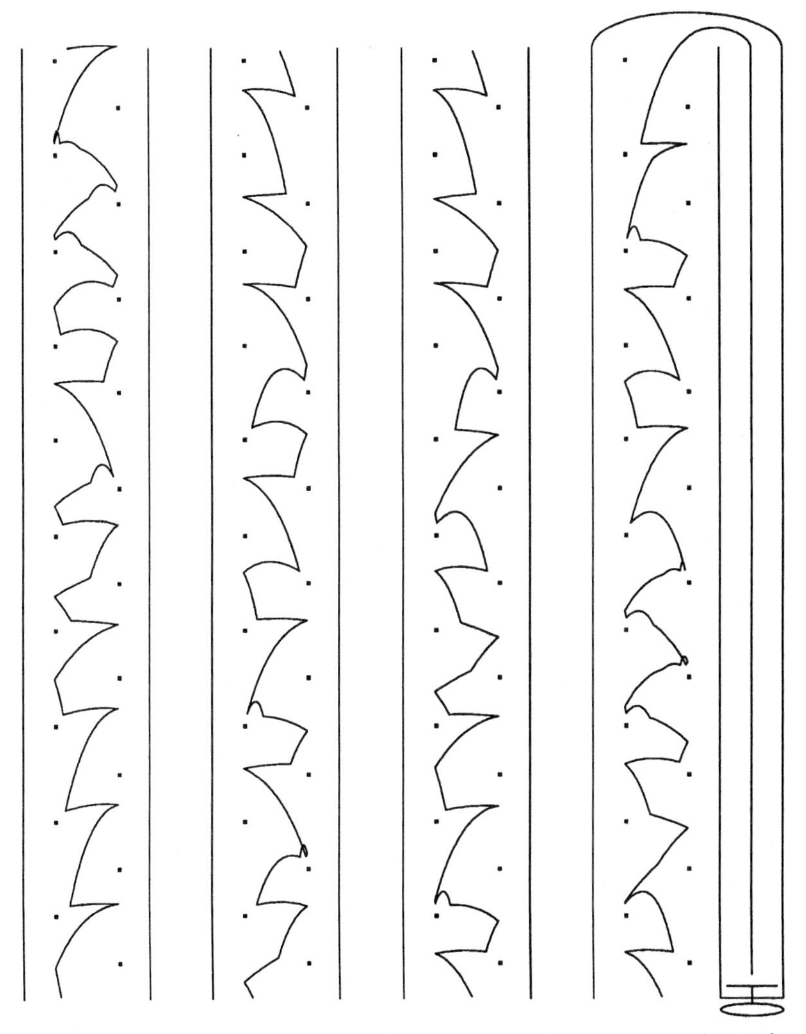

Figure 3. The elongated pinball machine, with the path of the center of a single ball. For display the machine has been divided into four columns, with columns to the left following those to the right. The vertical scale has been compressed; hence a ball striking a pin from above appears to be closer than one striking from the side.

riodicity will fail to become established, because even if, say, the tenth occurrence nearly repeats the ninth, and the ninth nearly repeats the eighth, etc., the tenth need not be particularly close to the first. In a case like this, the system is called *almost periodic*. If, then, some other compact system fails to exhibit periodicity or almost-periodicity, it presumably is chaotic.

In the case of Figure 3, we see that, if the behavior were not chaotic, the small differences between the early parts of the third and second columns would not amplify, and the remainder of the third column would look almost like the remainder of the second, while the fourth column would look almost like the third. Actually, however, the ball strikes the first pin in the fourth column and misses the second, while just the opposite happens in the third column. The fact that the fourth column does not repeat the third, or more generally that the path proves to be neither periodic nor almost periodic, is therefore sufficient evidence that the ball is behaving chaotically.

Returning to the flapping flag, we see that we may not need to photograph it, or go through other involved procedures that might reveal chaotic behavior. If near repetitions really do tend to occur after about ten seconds rather than an hour or longer, our simplest procedure may be just to listen to the continual puffs and plops, and to note whether they have a regular rhythm or whether they seem to occur "at random." I suspect, in fact, that the popular conception of something acting randomly is something varying with no discernible pattern, rather than something bearing the less easily detected property of sensitive dependence. Indeed, absence of periodicity has sometimes been used instead of sensitive dependence as a definition of chaos. Note, however, that if a system is not compact, so that close repetitions need never occur, lack of periodicity does not guarantee that sensitive dependence is present.

Zeroing In on Chaos

"Chaos," as a standard term for nonperiodic behavior, seems to have received its big boost in 1975 with the appearance of a now widely quoted paper by Tien Yien Li and James Yorke of the University of Maryland, bearing the terse title "Period Three Implies Chaos." In a mapping, a sequence of *period three* is one in which each state is identical with the state that occurred three steps earlier, but not with the state that occurred one step or two steps earlier; sequences with other periods are defined analogously. The authors showed that, for a certain class of difference equations, the existence of a single solution of period three implies the existence of an infinite collection of periodic solutions, in which every possible period—periods 1, 2, 3, 4, . . . —is represented, and also an infinite collection of nonperiodic solutions. This situation, in which virtually any type of behavior may develop, seems to fit the nontechnical definition of chaos, and it is not obvious that Li and Yorke intended to introduce a new technical term.

They might as well have done so. In the ensuing years the term has appeared with increasing frequency, and when, in 1987, it became the key word in the title of James Gleick's popular book, its permanence was virtually assured.

In the process of establishing itself as a scientific term, "chaos" also picked up a somewhat different meaning. Li and Yorke had used the term when referring to systems of equations that possess at least a few nonperiodic solutions, even when most solutions may be periodic. In systems that are now called chaotic, most initial states are followed by nonperiodic behavior, and only a special few lead to periodicity. I shall refer to chaos in the sense of Li and Yorke as *limited chaos,* calling chaos *full chaos* when it is necessary to distinguish it from the limited type.

You may wonder in what sense "most" solutions can behave in one manner when the "few" that behave otherwise are clearly infinite in number. Actually, occurrences of this sort are rather common. Consider, for example, a square and one of its diagonals. The number of distinct points on the diagonal is infinite, but, in an obvious sense, most points within the square lie off the diagonal.

Suppose that you decide to use the square as a dartboard, and that you have a dart whose point is infinitely sharp, so that there are an infinite number of different points that you might hit. If your aim is merely good enough to make it unlikely that you will miss the square altogether, your chances of striking a rather narrow band surrounding the diagonal are rather low, while your chances of striking a much narrower band are much lower, and the probability that you will hit a point exactly on the diagonal is smaller than any positive number that you can name. Mathematicians would say that the probability is zero. Clearly, however, a zero probability is not the same thing as an impossibility; you are just as likely to hit any particular point on the diagonal as any particular point elsewhere.

In limited chaos, encountering nonperiodic behavior is analogous to striking a point on the diagonal of the square; although it is possible, its probability is zero. In full chaos, the probability of encountering *periodic* behavior is zero.

There is a related phenomenon that, unlike full chaos or limited chaos, has long been familiar to almost everyone, even if not by its technical name, *unstable equilibrium.* If you have ever tried to make a well-sharpened pencil stand on its point, you will in all likelihood have found that less than a second will elapse between the moment when you let go of the pencil and the time when it strikes the table. Simple calculations indicate that a pencil is several million times less likely to stand up for two seconds than for one, and several million million times less likely to stand up for three seconds than for one, and that, in fact, if every human being who ever lived had devoted his or her entire life to attempting to make sharpened pencils stand on end, it would be highly unlikely that even one pencil would have stood up for six seconds. Of course, you may have somewhat better luck if the points of your pencils become slightly worn.

Nevertheless, theory indicates that a pencil standing exactly vertically will continue to stand forever; it is in equilibrium. A state of *equilibrium* is one that remains unchanged as time advances. An equilibrium is *unstable* if a state that differs slightly from it, such as one that you might purposely produce by disturbing it a bit, will presently evolve into a vastly different state—a fallen pencil instead of a standing one. It is *stable* if a slight initial disturbance fails to have a large subsequent effect. The concept of equilibrium, stable or unstable, can be extended to include periodic behavior.

The vertical pencil is typical of systems in unstable equilibrium. The reason that in practice a pencil cannot be made to stand on its point is that neither the hand nor the eye can distinguish between a pencil that is truly vertical and one that is tilted by perhaps a tenth of a degree, and the mathematical probability of picking the one vertical state from among the infinitely many that seem vertical is zero. Moreover, in the real world, even if unstable equilibrium could be precisely achieved, something would soon disturb it.

The definition of unstable equilibrium has much in common with that of sensitive dependence—both involve the amplification of initially small differences. The distinction between a system that merely possesses *some* states of unstable equilibrium and one that is chaotic is that, in a system of the latter type, the future course of every state, regardless of whether it is a state of equilibrium, will differ more and more from the future courses of slightly different states.

Chaotic systems may possess states of equilibrium, which are necessarily unstable. In the pinball machine such a state will occur if the ball comes to rest against a pin, directly upslope from it. A slight disturbance will cause the ball to roll away to the left or right.

It may seem that in seeking bigger and better pinball machines I have overlooked the simplest example of all. Consider an object moving without friction on a horizontal plane—effectively a pinball machine with no pins, no slope, and no walls. The object will move straight ahead, at a constant speed. If we should alter its speed or direction by even the minutest amount, its position will eventually be far from where it would have been without the alteration.

Should we call this chaos? Despite the sensitive dependence, there is no irregularity or seemingly random behavior. All possible paths are straight lines, expressible by simple mathematical formulas. Instead of recognizing a form of chaos with radically different properties, it would seem logical to conclude that our original definition in terms of sensitive dependence lacks some essential qualification.

Although there are a number of procedures for making the definition more acceptable, one approach is particularly simple. The object sliding without friction is an example of a dynamical system in which certain quantities may assume any values initially, but will then retain these values forever; these quantities are really constants of the system. In the present example, the *velocity* remains constant. If two nearby objects have identical velocities—speeds and directions—and differ in their initial states only by virtue of differing in their positions, they will not move apart.

Let us therefore amend our definition of chaos. First, for this particular purpose, let us refer to any quantity whose value remains unaltered when a system evolves without our interference, but may be altered if we introduce new initial conditions, as a *virtual constant.* For an object sliding without friction on a horizontal surface, the speed and direction are virtual constants; for one that slides without friction inside a bowl, the speed and direction vary, but the total energy is a virtual constant. Next, let us distinguish between changes in initial conditions that *do* alter the value of at least one virtual constant and those that do not, by calling the former changes *exterior* and the latter ones *interior.*

We may now redefine a chaotic system as one that is sensitively dependent on *interior* changes in initial conditions. Sensitivity to exterior changes will not by itself imply chaos. Concurrently, we may wish to modify our idea as to what constitutes a single dynamical system, and decide that, if we *have* altered the value of any virtual constant, we have replaced our system by another system. In that case chaos, as just redefined, will be equivalent to sensitive dependence on changes that are made within one and the same system. For systems without virtual constants, such as the elongated pinball machine, all changes are necessarily interior changes, and the modified definition is the same as the previous one. For many other systems, including various objects that move without friction, the modified definition—sensitivity to interior changes—will lead to more acceptable conclusions.

Have we been glimpsing a phenomenon that can arouse enough interest to become the subject of an extensive scientific theory? Probably few people care whether a flag flaps chaotically or rhythmically. Chaos in the pinball game is important, and frustrating, to anyone who is seriously trying to win. Chaos in the heartbeat, when it occurs, is of concern to all of us.

CHAPTER 2

A Journey into Chaos

Chaos in Action

THE PINBALL MACHINE is one of those rare dynamical systems whose chaotic nature we can deduce by pure qualitative reasoning, with fair confidence that we have not wandered astray. Nevertheless, the angles in the paths of the balls that are introduced whenever a ball strikes a pin and rebounds—rather prominent features of Figure 1—render the system somewhat inconvenient for detailed quantitative study. For an everyday system that will vary more smoothly, and can more easily serve to illustrate many of the basic properties of chaotic behavior, even though it may not yield so readily to descriptive arguments, let us consider one that still bears some resemblance to the pinball machine. The new system will again be a slope, with a ball or some other object rolling or sliding down it, but there will be no pins or other obstacles to block a smooth descent, and the slope may be of any size.

We should not expect to encounter chaos if our slope is a simple tilted plane, unless our object is an elliptical billiard ball of *The Mikado* fame, so let us consider a slope with a generous scattering of smoothly rounded humps. These may even be arranged like the pins in some pinball machine—perhaps the machine appearing in Figure 1—but they will not have the same effect; even though an object approaching a hump obliquely may be deflected about as much as a ball glancing off a pin, an object that encounters a hump straight ahead can travel smoothly over the top instead of rebounding. We wish to discover a system of this description that behaves chaotically, if, indeed, there is one.

At this point we have the usual options. We can make an excursion to the country, and seek a slope with plenty of humps, but without any obstacles, such as trees or boulders, that might play the role of the pins in the machine. We can then let a soccer ball roll down and observe its motion, but an irregularly meandering path may not indicate sensitive dependence, since the humps, and hence the deflections that they produce, may not be regularly spaced. Instead of detecting chaos by its lack of rhythm, we would have to look for sensitive dependence directly. To do this, we could release another or preferably several other balls from the same point to see whether they follow different paths, hoping that any observed divergence will not be the product of an unobserved gust of wind. We could virtually remove the wind problem by using bowling balls, but we might be disinclined to carry several of them up the slope.

We may therefore prefer to turn to the laboratory and build our own slope, making every hump just like every other one and spacing them uniformly. The slope can sit on a table top, and the rolling object can be a marble. Like the soccer ball, the marble will be deflected by the humps, but now there will be nothing to prevent its motion from varying periodically. If it follows an irregular path anyway, we should suspect the presence of chaos. Again, we might choose to compare the paths of several marbles.

If we are mathematically inclined, we may wish to move from the laboratory to the computer. Instead of physically building a slope, we can choose a mathematical formula for its topography. Instead of watching an object as it moves down the slope, we can solve the equations that describe the manner in which its motion will vary. Let us examine in detail what can happen if our envisioned slope resembles the one pictured in Figure 4, which shows an oblique view of a cut-out section. The humps may remind some of us of moguls on a ski slope, and, although a computer program can handle slopes of any size with equal ease, I shall choose dimensions that are fairly typical of ski slopes. Since some of you may live far from ski territory, I should add that these moguls are not eastern rulers nor industrial giants who have ventured onto the snow. Compare Figure 5, where you can perhaps just detect a lone skier tackling the ubiquitous moguls on the Cat's Meow, a challenging slope at Loveland Basin ski area in Colorado. The moguls seem rather regularly spaced, though not enough so to fit a simple mathematical formula.

Since our object will travel continuously rather than in jumps, our equations will be differential equations. They will be statements, in mathematical symbols, of one of Isaac Newton's laws of motion, which simply equates the acceleration of any particle of matter—the rate at

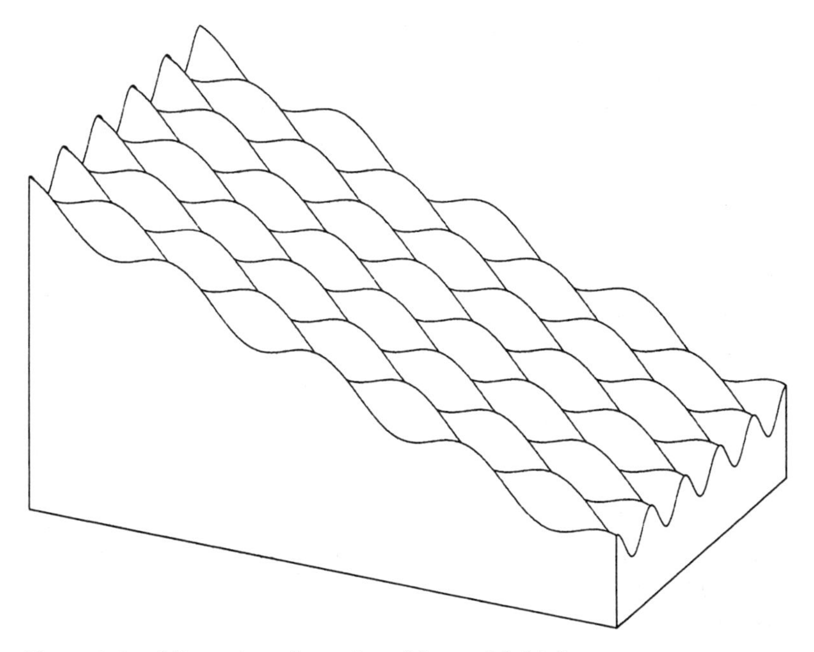

Figure 4. An oblique view of a section of the model ski slope.

which its velocity is changing—to the sum of the forces acting on the particle, divided by the mass of the particle. A complicated object can be treated as an aggregation of particles. The equations appear in a mathematical excursion in Appendix 2.

A ball or some other object rolling down a slope would acquire considerable spin, and to avoid the resulting mathematical complications I shall assume that our object is one that will slide. It might be a flat candy bar that has fallen from a skier's pocket, or even a loose snowboard, but not an animate skier, and I shall simply call it a board. Its motion will be controlled by the combined action of three forces. One of these is the pull of gravity, directed vertically downward. Another is the resistance of friction, directed against the velocity. Finally there is the force that the slope exerts against the board, directed at right angles to the slope, and opposing the effect of gravity to just the extent needed to keep the board sliding instead of sinking into the slope or taking off into the air.

Laboratory models are of necessity real physical systems, even though they need not be like any encountered in nature, but mathematical models seldom duplicate any concrete systems exactly. Real boards may be flexible, and their orientations may vary, but in our ski-slope model we shall disregard these possibilities, and treat the board as if it were a single particle. It will be convenient to choose an oversimplified law of friction, letting the resistance be directly proportional to the speed of the board—doubling the speed will double the resistance. Stated otherwise, if the *damping time* is defined as the ratio of the speed to the rate at which friction is lowering the speed, it will be convenient to let the damping time be constant. Its reciprocal, the *coefficient of friction*, will therefore be constant. It will then be easy to formulate a system of differential equations whose solutions will reveal how the board will move.

Our model would be more realistic if the resistance of friction were made to vary also with the force of the slope against the board, so that, for example, when the board is nearly taking off, presumably because it is shooting over a mogul, the frictional effect will be greatly reduced. Wind resistance should also be modeled, but these refinements are unlikely to have much qualitative effect.

There are different rules for laboratory and computer studies. We tend to associate laboratory experiments with high precision, but in some instances it is possible to omit some measurements altogether and still obtain an answer. In a computer experiment there is no way to begin until numerical values have been assigned to every relevant quantity.

Our model contains a number of constants. We therefore have the option of working with any one of an infinite number of dynamical systems, since, as we have noted, whenever we alter the value of any constant we produce a new system, perhaps with a new typical behavior—observe how oiling an aging machine, and thereby lowering its coefficient of friction, can pep up its performance. Systems that are formally alike except for the values of one or more constants are said to be members of a *family* of dynamical systems. Often we refer to a whole family as simply a dynamical system, when we can do so without causing confusion.

We should anticipate the possibility that only certain combinations, if any, of values of our constants—the average pitch of the slope, the height and spacing of the moguls, and the coefficient of friction—will lead to chaos. Gravity is also an essential constant, but its magnitude has already been determined for us as long as we keep our experiment earthbound. In contrast to the pinball model, where chaos is virtually assured, there is no rule for determining in advance of our computa-

Figure 5. Moguls on a real-world ski slope.

tions, in the new model, what combinations of values for the constants will work. We must discover them by trial and error, possibly obtaining suitable values on the first try, possibly encountering them only after many tries, and possibly not finding them at all and concluding, perhaps incorrectly, that the model will never produce chaos.

Since I have introduced the ski slope for the purpose of illustrating the fundamental properties of chaos, I shall choose values from a successful try. For convenience, let the slope face toward the south. Let its average vertical drop be 1 meter for every 4 meters southward. Imagine a huge checkerboard drawn on the slope, with "squares" 2 meters wide and 5 meters long, and let the centers of the moguls be located at the centers of the dark squares, as illustrated in Figure 6, which also shows a possible path of a board down the slope. Let the damping time be 2 seconds. As detailed in Appendix 2, the formula selected for the topography of the slope will place pits in the light squares of the checkerboard, as if snow had been dug from them to build the moguls. Let each mogul rise 1 meter above the pits directly to its west and east. Like the pinball machine, the ski slope would have to be infinitely long to afford a perfect example of chaos.

We shall be encountering references to the four variables of the model so often that I shall give them concise names. These might as well consist of single letters, and they might as well be letters that have served as mathematical symbols in the differential equations, or in the computer programs for solving them. Let the southward and eastward distances of the board from some reference point, say the center of a particular pit, be called X and Y, respectively, and let the southward and eastward components of the velocity—the rates at which X and Y are currently increasing—be called U and V. Note that when we view the slope directly from the west side, as we nearly do in Figure 4, or as we can do by giving Figure 6 a quarter turn counterclockwise, X and Y become conventional rectangular coordinates. Alternatively, but with some loss of mathematical convenience, we could have chosen the board's distance and direction from the reference point, and its speed and direction of motion, as the four variables.

What I have been calling the centers of the moguls or the pits are actually the points where the slope extends farthest above or below a simple tilted plane. These are the centers of the checkerboard squares. The actual highest point in a dark square is about 1.5 meters north of the center

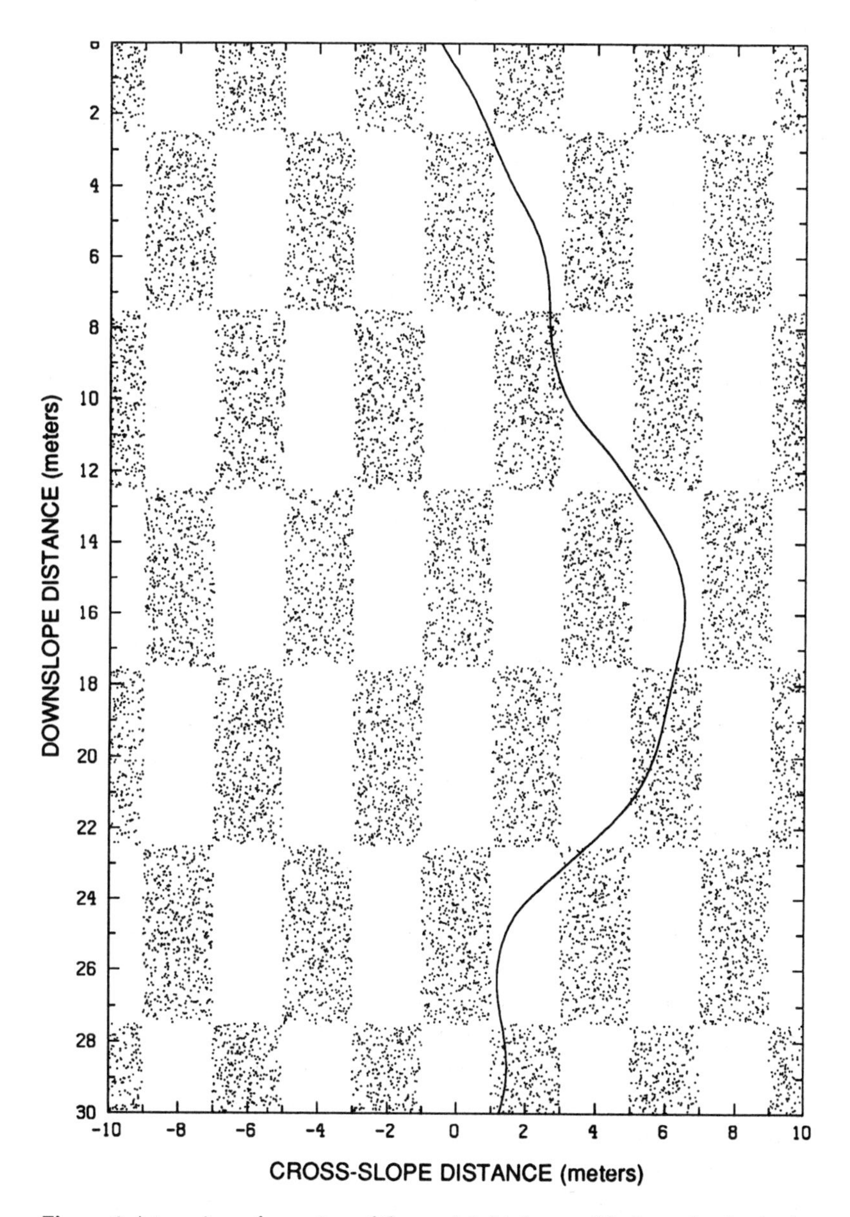

Figure 6. A top view of a section of the model ski slope, with the path of a single board sliding down it. The shaded rectangular areas of the slope project above a simple inclined plane, while the unshaded areas project below.

of the mogul, while the lowest point in a light square is a like distance south of the center of the pit, and I shall call these points the high points and low points. They may be detected on the forward edge of the section of the slope in Figure 4.

To begin a computation, we can give the computer four numbers, which specify numerical values of *X* and *Y*, say in meters, and *U* and *V*, say in meters per second. In due time the computer will present us with more numbers, which specify the values of the same variables at any desired later times. As our first example, let *X*,*Y*, *U*, and *V* be 0.0, -0.5, 4.0, and 2.0, implying that the board starts half a meter due west of the center of a pit, and heads approximately south-southeastward. The board will then follow the sample path shown in Figure 6.

To see whether the board descends chaotically, let us turn to Figure 7, which shows the paths of seven boards, including the one appearing in Figure 6, as they travel 30 meters southward. All start from the same west-east starting line with the same speed and direction, but at points at 10-centimeter intervals, from 0.8 to 0.2 meters west of a pit. A tendency to be deflected away from the moguls is evident. The paths soon intersect, but the states are not alike, since now the boards are heading in different directions, and soon afterward the paths become farther apart than at first. By 10 meters from the starting line, the original 0.6-meter spread has more than doubled, and by 25 meters it has increased more than tenfold. Clearly the paths are sensitively dependent on their starting points, and the motion is chaotic.

As already noted, an essential property of chaotic behavior is that nearby states will eventually diverge no matter how small the initial differences may be. In Figure 8, we let the seven boards travel 60 meters down the slope, starting from points spaced only a millimeter apart, from 0.503 to 0.497 meters west of the pit. At first the separation cannot be resolved by the picture, but by 30 meters it is easily detectable, and subsequently it grows as rapidly as in Figure 7. The ski slope has passed another critical test. The initial separation can be as small as we wish, provided that the slope is long enough.

Lest the paths remind some of you of ski tracks, I should hasten to add that they do not follow routes that you, as knowledgeable skiers, would ordinarily choose. They might approximate paths that you would take if you fell and continued to slide.

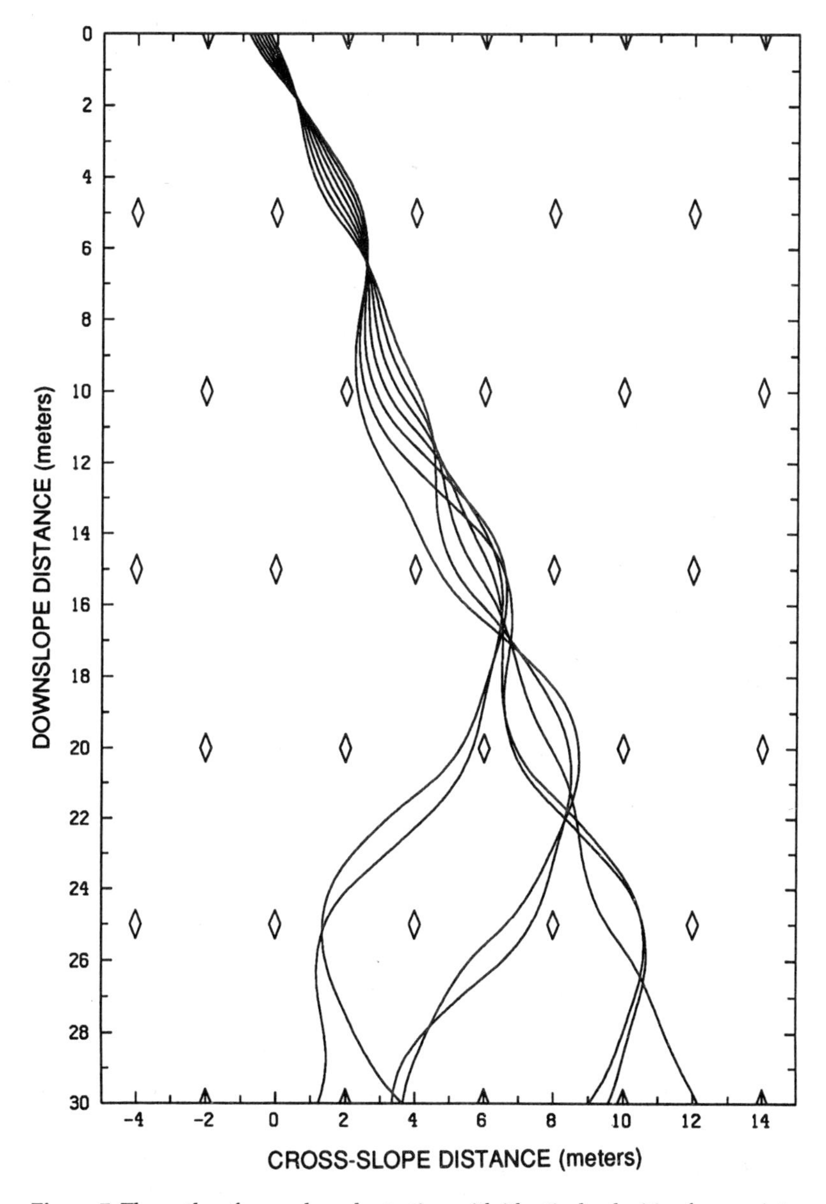

Figure 7. The paths of seven boards starting with identical velocities from points spaced at 10-centimeter intervals along a west-east line. The small diamonds indicate the locations of the centers of the moguls.

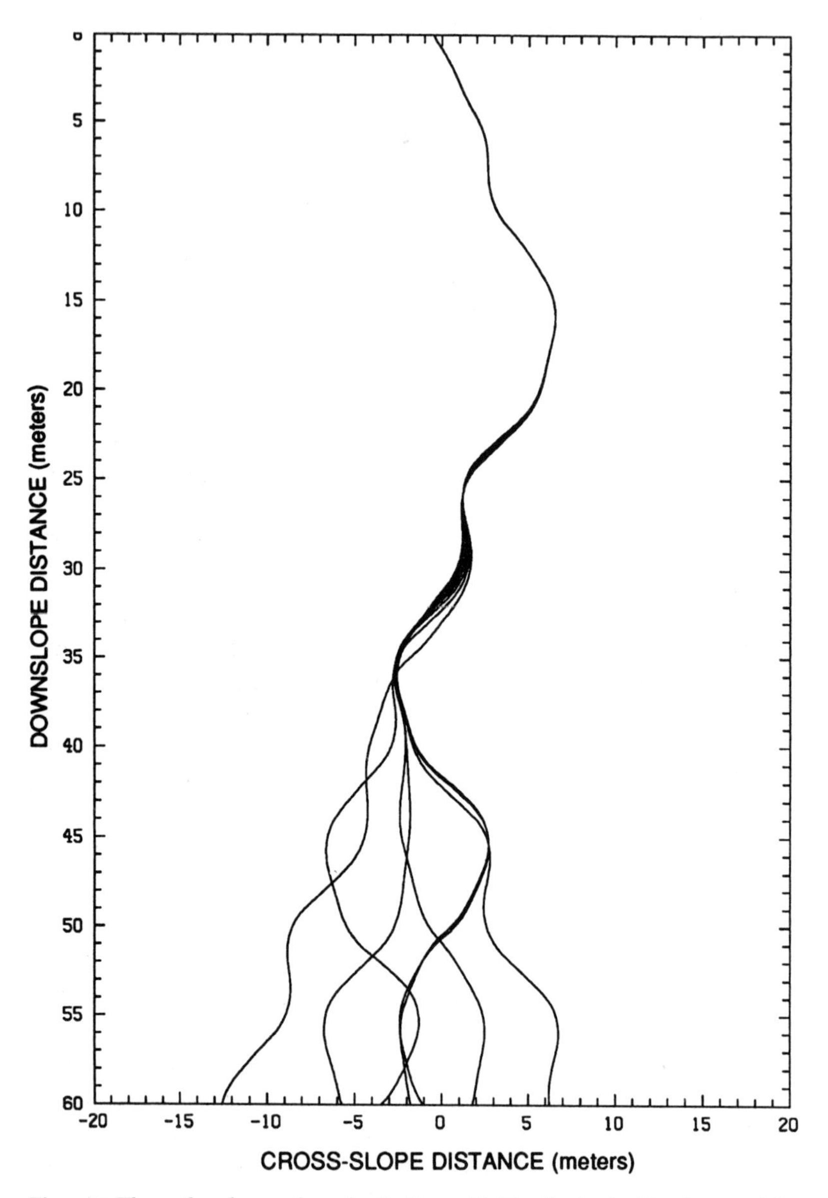

Figure 8. The paths of seven boards, starting with identical velocities from points spaced at 1-millimeter intervals along a west-east line.

The reader who looks at things as they are in nature need not be perplexed by what is meant by displacing a snowboard, not to mention a fallen skier, by 1 millimeter. We have been looking at a mathematical model, which, like most models, ignores something. In this case it ignores the size of the sliding object, which, presumably quite unrealistically, is assumed to slide just as it would if it were no larger than a pinhead, and if the slope were so smooth that a pinhead could continue to slide. It would be quite possible to write a new computer program, which could deal with sliding objects of various sizes and shapes, but which would have to be based on a more involved mathematical analysis.

For the alternative means of detecting chaos—directly observing a lack of periodicity—let us turn to Figure 9, which extends the original path to 600 meters down the slope. The board shows a noticeable tendency to wiggle about lines that pass either southeastward or southwestward between the moguls, but the switches from one direction to the other show no sign of occurring at regular intervals. Let us see what we can infer from this behavior.

First, we are interested in periodicity, or the lack of it, as time increases, but Figure 9 shows only what happens as downslope distance, i.e., *X*, increases. We should note, then, that since the board always moves down the slope and its speed does not vary too greatly, distance can serve as an approximate measure of time, with 3.5 meters equaling about 1 second. Measuring time in this manner is like using a clock that sometimes runs a bit too fast or too slow, but always runs forward.

If the whole system varied periodically with time, the speed of the clock, which is certainly a feature of the system, would have to vary periodically, with the same period, and the variations of the system, as measured by this clock, would likewise be periodic. Failure to vary periodically with downslope distance therefore implies absence of periodicity in real time.

Next, like the pinball machine, the slope with the board coasting down it is not a compact system, since *X* must increase without limit, and *Y* may do likewise. A simple change of variables, however, will produce a system that is mathematically compact. Note that it is possible to change the variables of a tangible system without altering the system itself. We would do this, for example, if we decided to express a velocity

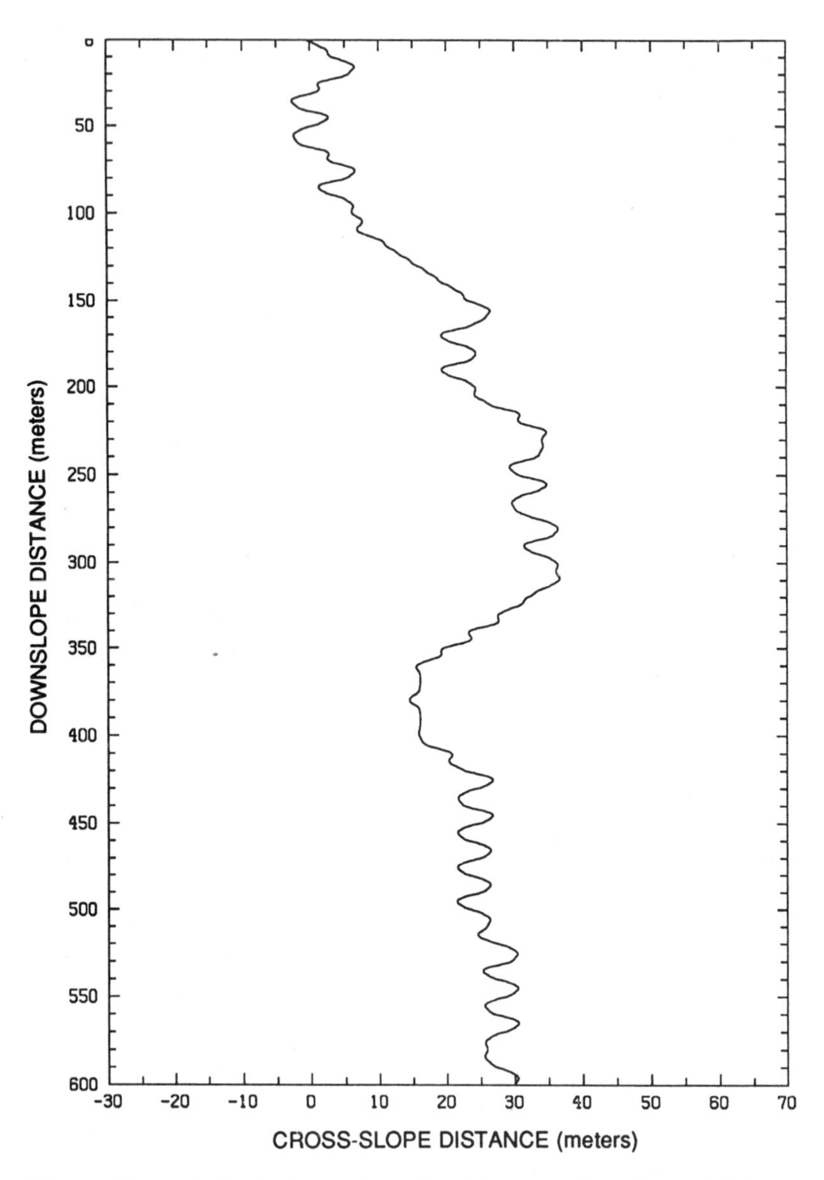

Figure 9. The path of a single board traveling 600 meters down the model slope. Note that the north-south scale has been compressed, as if we were looking at the slope from beyond the base.

in terms of speed and direction instead of southward and eastward components.

In the present case, we first partition the slope into 5-meter by 4-meter rectangles. Each rectangle contains an entire light square of the checkerboard and extends halfway through the dark squares to the west and east, so that the center of a pit lies at the center of each rectangle, and centers of moguls lie at the midpoints of the west and east sides. Then, using lower-case instead of upper-case names, we let x and y, the new variables, be the southward and eastward distances of the board from the center of the rectangle that it currently occupies. Most of the time x will increase continuously, just as X does, but x will jump from 2.5 back to -2.5 whenever the board enters the next rectangle down the slope, at which time y will abruptly increase or decrease by 2 meters, according to whether the board has left the previous rectangle nearer to its southwest or its southeast corner. Also, y will jump from 2.0 to -2.0, or from -2.0 to 2.0, when the board crosses one side of a rectangle or the other, but U and V will retain their original meanings as components of velocity. Since the slope has the same shape in each rectangle, the values of x, y, U, and V at any future time can be determined from the present values without knowing X and Y, i.e., without knowing which pit the board is near. Since, like U and V, x and y vary only over limited ranges, the new system is compact.

We have therefore identified a compact mathematical system in which U and V can vary exactly as they do in Figure 9. In this figure, positive and negative values of V show up respectively as southeastward and southwestward progressions. Since these do not appear to alternate in any regular sequence, at least one variable, V, is patently not varying periodically, and we can conclude fairly safely, even without reference to Figure 7 or 8, that the behavior of the board is chaotic. The only alternative would be the unlikely one that it is periodic with a period exceeding 600 meters in X, or about three minutes—too long to be revealed by the figure.

The path down the slope in Figure 9 is actually a graph of Y against X; with a quarter turn it will look like a conventional graph. In effect the board draws its own graph as it slides down the slope. Many of the fundamental concepts in dynamical-systems theory can be represented by graphs, but these graphs are not always plots of values of a single variable against values of one other. Often we are interested in the simultaneous values of several variables, and we may wish that we could have graphs in more than two dimensions, or even in as many dimensions as the number of variables in our system. If this number is large, it will be utterly impossible to construct some of the desired graphs, but, if it is fairly small, constructions in two or three dimensions can sometimes fulfill our needs.

As dynamical systems go, the ski slope is fairly simple, having only four variables; compare even a crude model of the flapping flag. Nevertheless, most persons do not find a curve or any other kind of graph in four-dimensional space particularly easy to visualize.

For example, consider a length of cord, perhaps a clothesline, with an overhand knot in it, and with its ends fastened to opposite walls of a room. In the three-dimensional world in which we live, we cannot remove the knot without detaching one end or mutilating the cord. Knot theorists have shown that in four dimensions a similar knot can be readily removed while the ends remain fastened; effectively it is not a knot at all. This finding does not seem unreasonable, but visually I do not find it at all obvious.

It would be convenient to work with a system with fewer than four variables. We can produce such a system by doctoring up our ski-slope model.

Physical complexity and mathematical complexity are not the same thing. It is quite possible to replace one system by another that is physically more complicated but mathematically simpler. In our new model we shall retain the original ski slope while replacing the board by a sled. The sled will be equipped with brakes but no steering mechanism. The driver is supposed to apply the brakes to just the extent needed to hold the downslope speed, *U*, constant, while the cross-slope speed, *V*, may continue to vary. If the moguls are rather high, a sled sliding over a mogul will slow down even if the brake is completely released, so in this case the sled must also be equipped with a motor, and the driver must use the accelerator when the need arises.

The driver, who is unlikely to appreciate his lack of control over the sled's direction and may not find much comfort in the knowledge that his ride may be chaotic, will probably need considerable practice before he can hold the southward speed even approximately constant. He may well opt for some special electronic equipment, which will observe the slope just ahead and apply the brakes or accelerator accordingly. Certainly a properly working sled of this sort would be physically much more complicated than a simple board. Designing a laboratory model of the sled would involve similar complications, and might require similar electronic equipment. In contrast, a computer program for the new system will be *simpler* than the original program, but the big advantage will not be the modest saving in computer time but the fact that one original variable, the southward speed U, has become a constant, still called U, so that the new system has only three variables. These may be chosen as X, Y, and V, or instead as x, y, and V. Moreover, X, the downslope distance, now increases uniformly with time, and the clock that used to run too fast or too slow now works perfectly.

As with the original system, we can expect that not all combinations of values of the constants will lead to chaos. Although our constants now include U, the downslope speed, they no longer include the coefficient of friction, whose value is controlled by the driver when he brakes, and hence mathematically is determined by X, Y, and V. Use of the accelerator will appear in the program as negative friction.

For our example we shall let U be 3.5 meters per second—a value close to the average value of U in the original example—while the remaining constants will retain their former values. Figure 10 shows what happens when seven sleds, spaced at 10-centimeter intervals along the same west-east starting line, start with equal velocities. Although the details differ from those in Figure 7, the message is the same; the system is chaotic.

The Heart of Chaos

An expression that has joined "chaos" in working its way into the scientific vocabulary, and that has aroused a fair amount of popular interest as well, is "strange attractor." Let us see what is meant by an attractor, and what one must be like to be considered strange.

If we take a look at some real-world phenomenon that has caught our attention, we are likely to find that certain conceivable modes of behavior simply do not occur. A pendulum in a clock in good working order will not swing gently at times and violently at others; every swing will look like every other one. A flag in a steady breeze will never hang limp, nor will it extend itself directly into the breeze, no matter how long we wait. Subfreezing temperatures will not occur in Honolulu, nor will rela-

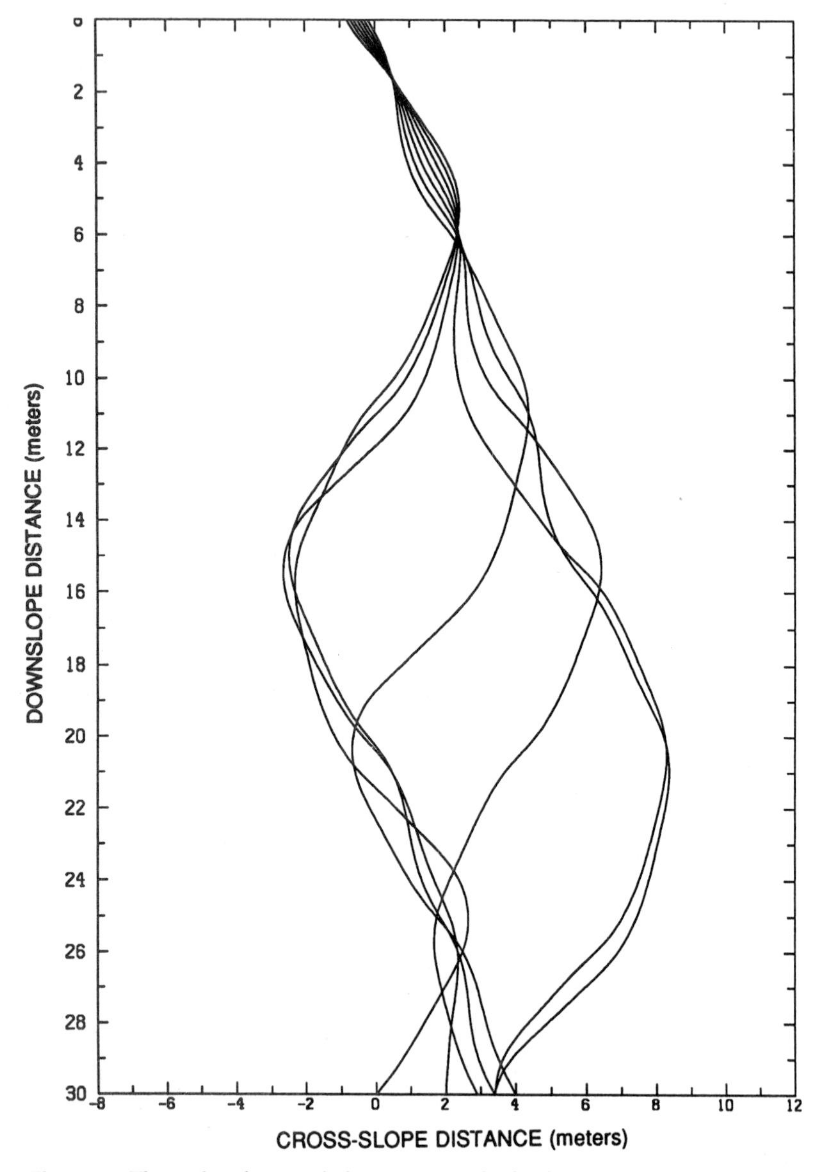

Figure 10. The paths of seven sleds, starting with identical velocities from points spaced at 10-centimeter intervals along a west-east line.

tive humidities of 15 percent. The states of any system that do occur again and again, or are approximated again and again, more and more closely, therefore belong to a rather restricted set. This is the set of *attractors*.

When we perform a numerical experiment with a mathematical model, the same situation arises. We are free to choose any meaningful numbers as initial values of the variables, but after a while certain numbers or combinations of numbers may fail to appear. For the sled on the slope, with the value of U, the downslope speed, fixed at 3.5 meters per second, we can choose any initial value for V, the cross-slope speed, and any values for x and y within restricted ranges. Computations show that V will soon become restricted also, remaining between -5.0 and +5.0 meters per second.

Moreover, even the values of V that continue to occur will not do so in combination with certain values of the other variables. The sled will frequently slide almost directly over a mogul, and it will often move almost directly from the northwest or northeast, but whenever it is crossing a mogul it will be moving only from almost due north. To the computer, this means that, if x is close to 0.0 and y is close to -2.0 or +2.0, V will prove to be close to 0.0. Again, the states that do manage to occur, after the disappearance of any transient effects that may have been introduced by the choice of initial conditions, will form the set of attractors.

The interest in attractors that has accompanied and perhaps stimulated the recent surge of interest in chaos has been partly due to the striking appearance of certain ones—the "strange" ones. Can a set of attractors, which is simply a collection of states, have any appearance at all, let alone a strange one, particularly if the variables happen to be invisible quantities, like atmospheric temperature and pressure? It can indeed, if by an attractor we mean a graphical representation of an attractor.

As we saw earlier, we may sometimes wish that we could draw graphs or other diagrams in a space that has as many dimensions as the number of variables in our system. Often such a task is impossible, but even then the concept of these diagrams can be useful. The hypothetical multidimensional space in which such a diagram would have to be drawn is known as *phase space*.

In phase space, each point represents a particular state of a dynamical system. The coordinates of the point—distances in mutually perpendicular directions from some reference point, called the *origin*—are numerically equal to the values that the variables assume when the state occurs. A particular solution of the equations of a system—that is, the set of states following and perhaps preceding a particular initial state—is represented by a curve, often called an *orbit*, if the system is a flow. It is represented by a chain of points, also called an orbit, if the system is a mapping. An attractor may be represented by a simple or complicated geometrical structure.

In the minds of many investigators, an attractor and its pictorial representation in phase space are one and the same thing. In their terminology, a point *means* a state, and an orbit *means* a chronological sequence of states, so that a set of attractors can *be* a collection of points. When this collection consists of a single agglomeration, it is also *the* attractor. When it is composed of several unconnected pieces, and when no orbit passes from one piece to another, each piece is a separate attractor.

An attractor can sometimes be a single point. A pendulum in a clock that is not wound up will eventually come to rest, hanging vertically, regardless of how it has initially been set in motion. Since the state of the pendulum, before it comes to rest, can be described by two variables, the phase space of the pendulum must be two-dimensional; it must be a plane. On this plane, we may choose the horizontal distance of a point from the origin to equal the horizontal displacement of the bob from the low point of the swing. The vertical distance from the origin will not represent the bob's vertical distance from anywhere; instead we may *choose* it to have the same numerical value as the speed of the bob— *positive* when the bob is moving to the right and *negative* when it moves to the left. The attractor of the unwound clock will then be a single point on the plane—the origin—representing the state of rest.

The attractor of a pendulum in a clock that is always kept wound will be a closed curve, resembling an ellipse. In fact, if we measure the speed of the bob in suitable units, perhaps miles per hour or perhaps centimeters per second, it can resemble a circle, with its center at the origin. Regardless of how rapidly we start the pendulum swinging, it will soon acquire its normal behavior, and then, as it swings toward the right, the point representing its state will move to the right along the upper or positive half of the circle, crossing the top when the bob attains its greatest speed. As the pendulum swings back toward the left, the point will move back along the lower or negative half of the circle, after which it will continue its clockwise circuits.

Because, in this and many other systems, extremely large values of the variables cannot occur, except as transient conditions, points in the set of attractors cannot be too far removed from the origin. This means that they will occupy a rather restricted central region. They will indeed form the heart of the dynamical system.

For the sled on the slope, it is convenient to use x, y, and V as coordinates in the three-dimensional phase space. The central region, containing the attractor, will then fit into a rectangular box—a box within which x extends only from -2.5 to +2.5 and y extends only from -2.0 to +2.0, since by definition x and y are limited to these ranges, while V extends only from -5.0 to +5.0, since larger cross-slope speeds do not occur, except transiently.

There are numerous sophisticated computer programs for constructing pictures of three-dimensional objects, although some of them may not work very well when the object is a complicated attractor. For our system, let us adopt the simple procedure of displaying cross sections of the box. The simplest of these sections are rectangles, parallel to one face or another. Mathematically, it is easiest to work with rectangles on which x is constant, and on which V is plotted against y. Horizontal and vertical distances from a central point of a rectangle can then equal values of eastward distance and eastward speed, respectively, just as they do in the phase space of the pendulum when the clock faces south. We may then draw a picture of the cross section of the attractor by determining, and marking on a chosen rectangle, the locations of a large number of points that lie not only on the rectangle but also on the attractor. By looking at several parallel cross sections of the attractor we may get a good idea of its three-dimensional structure.

To determine the desired points, we may begin by taking *any* large collection of points on one of the rectangles. Each point will represent the initial state of one sled. Let us choose the rectangle on which x equals 2.5, so that the sleds all start from a west-east line that passes midway between a pit and the mogul directly to its south. The upper left panel in Figure 11, which, unlike the previous figures, is a picture of phase space instead of something tangible, contains five thousand points chosen at random. These are supposed to be a sampling of *all* points on the rectangle, representing the initial states of all sleds starting with any east-

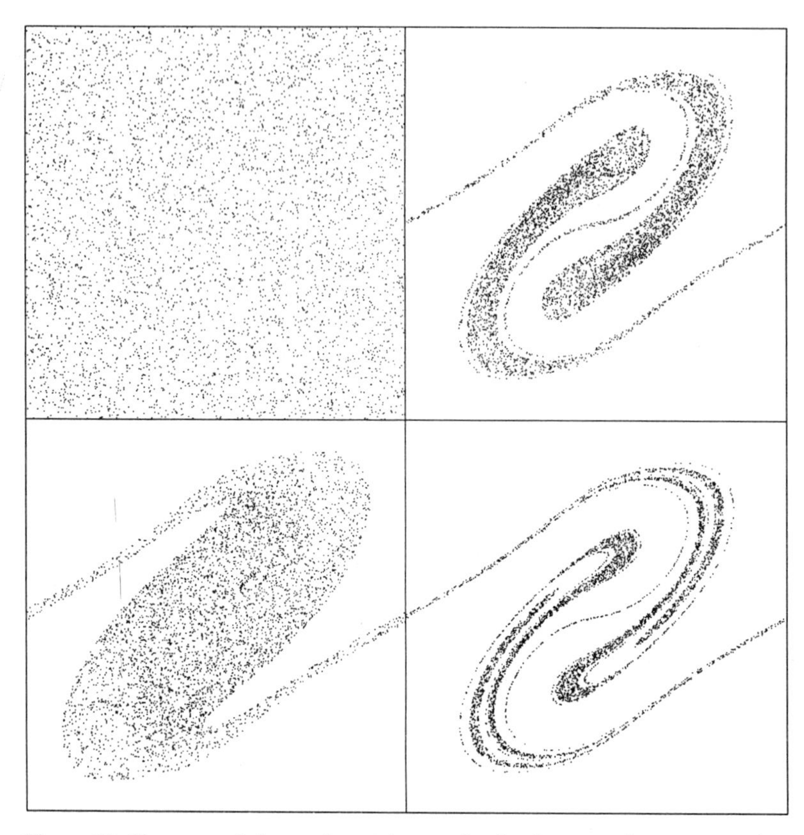

Figure 11. The upper left panel contains randomly chosen points representing the cross-slope positions and speeds of five-thousand sleds, all located on the same west-east line. The lower left and then the upper right and lower right panels represent the positions and speeds of the same sleds after they have traveled 5 and then 10 and 15 meters down the slope.

ward speed between -5.0 and +5.0 meters per second, from any point on the starting line. The pattern has no recognizable form; it is true chaos in the nontechnical sense of the word.

We next allow each sled to descend 5 meters, so that x equals 2.5 again. In the lower left panel we have plotted the five thousand points representing the newly acquired cross-slope positions and velocities. We find that the points have gathered themselves into a more or less elliptical region with two thin arms extending from it. There are large empty areas, representing states that cannot occur except transiently. Points on the left and right edges are confined to narrow bands near the midpoints of these edges, where V is close to zero; this implies that the sleds represented by these points are moving from nearly due north. Note that a horizontal line across the panel midway between the top and the bottom would have two stretches with no points at all.

The right-hand panels show what happens when the sleds have descended 10 and then 15 meters from the starting line. The points become attracted to regions that are more and more elongated and distorted; it is as if someone were continually twisting the central portion of the cross section clockwise, like the key in a wind-up toy, while holding the left and right extremities fixed. Each new set of points fits inside the preceding one. At 10 meters a line midway between the top and bottom of the panel has four empty stretches; at 15 meters it has eight.

The set to which these points will ultimately be attracted, if we continue the process, will be the cross section of the attractor. We can more or less see what it will look like by extrapolating from the panels in Figure 11. The assemblage of points will become infinitely elongated, infinitesimally thin, and infinitely distorted. Virtually any line crossing the region will have an infinite number of empty stretches, and between any two empty stretches there will be points, separated by still more empty stretches.

A more conventional way to determine a collection of points on the cross section of an attractor is to take a single initial point and plot a long chronological sequence, often omitting the first few points, which may represent transient conditions. In Figure 12 we see a plot of ten thousand consecutive values of eastward speed V against eastward distance y, occurring at 5-meter intervals of southward distance X, all with x, the southward distance from the nearest pit, equal to 2.5. Taking advantage of the symmetry of the attractor—compare Figure 11—I have added another ten thousand points by also plotting $-V$ against $-y$. Never mind that a ski slope 50 kilometers long, with its assumed pitch, would have to descend from higher than the summit of Mt. Everest, and would undoubtedly have more slippery snow at some elevations than at others; as usual we are working with a mathematical idealization. The figure is the cross section of the attractor—the one that is being approached by the successive panels in Figure 11. It clearly appears to be composed of many nearly parallel curves. The more heavily shaded curves represent sets of states that occur more frequently, while those that show up as open chains of points represent less common events.

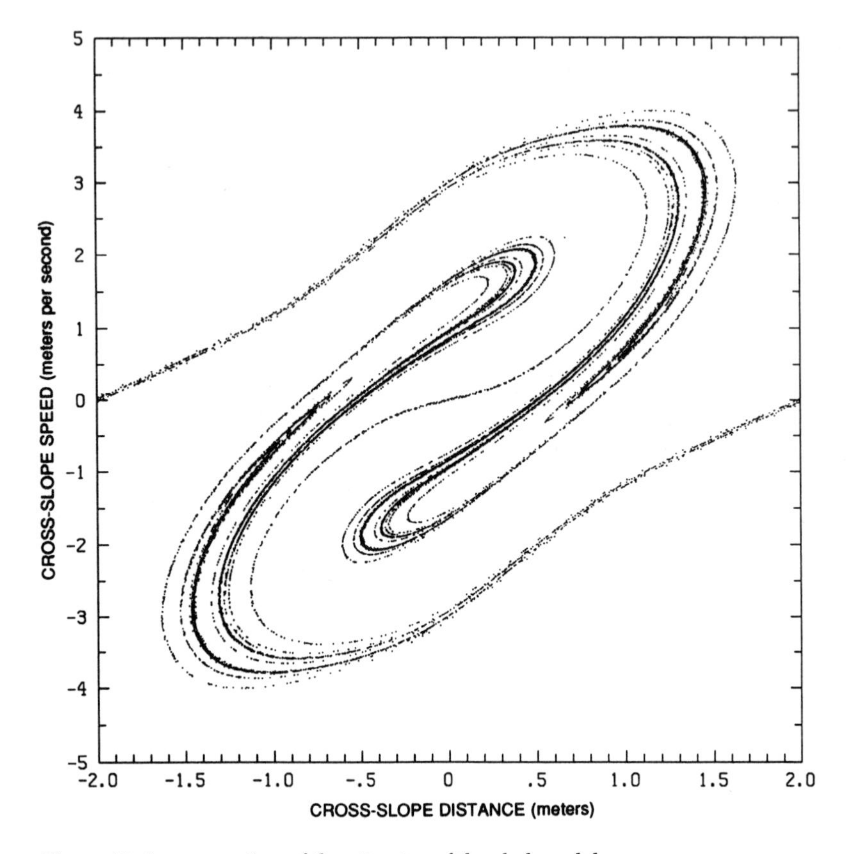

Figure 12. A cross section of the attractor of the sled model.

Like Figure 11, Figure 12 is not a picture of something spread out on the slope; it is a graph. Even though horizontal distance on the graph represents distance across the slope, vertical distance has little to do with distance up or down the slope. Alternative representations of the attractor would seem to be equally legitimate.

Since the points on the right-hand edge of the figure represent the same states as the points on the left-hand edge—they represent sleds directly south of a mogul—it would be logical to join them somehow, and we could do this by wrapping the figure around a cylinder, like the label on a soup can. Alternatively, we could take the paper on which the figure is drawn and roll it into a tube, with the figure on the inside, and then peer into the tube. What we would then see would look more or

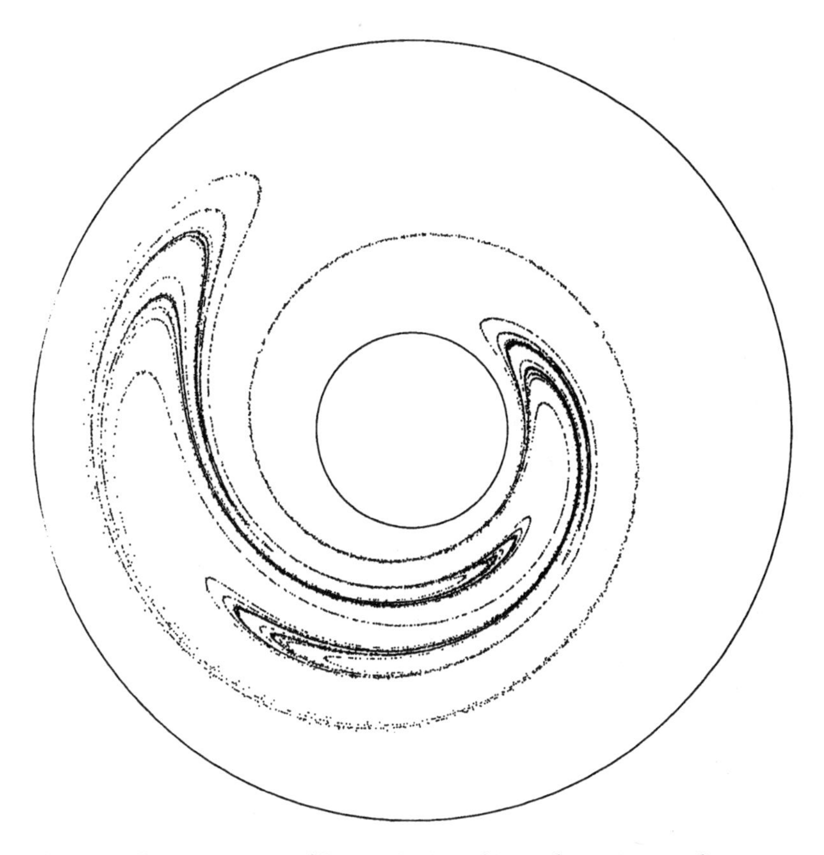

Figure 13. The cross section of Figure 12, viewed in an alternative coordinate system. The inner and outer circles correspond to the upper and lower boundaries of Figure 12. A line extending upward from the inner to the outer circle, not shown, would correspond to both side boundaries.

less like Figure 13—another picture of the same cross section of the attractor. The new form is more convenient for some purposes, and less so for others.

The system whose attractor we have been seeking is a flow, with a three-dimensional phase space; the sleds slide continuously along their paths, which have been determined by solving a system of differential equations. When we observe a sled only at 5-meter intervals down the slope, as we did in Figure 11, we are observing it only at special times, and so, just as when we decided to look at the pinball only when it was striking a pin, we are replacing a flow by a mapping derived from the flow. The mapping has a two-dimensional phase space, and the values of y and V at one of the special times completely determine the values at the next. We cannot write down the system of difference equations that relates the successive values of y and V, but we do not need to, since we have already obtained a numerical solution by solving the differential equations.

It follows that what we have been treating as a cross section of an attractor, and have displayed in Figures 12 and 13, is also a complete attractor—the attractor of the mapping. The process of converting a flow into a mapping with a lower-dimensional phase space was introduced by the French mathematician Henri Poincaré as a part of his novel approach to the problems of celestial mechanics. The cross sections, which he referred to as surfaces of section, are now called *Poincaré sections,* and the mappings produced by taking cross sections are called *Poincaré mappings.*

For a perspective view of the attractor in three dimensions, we turn to Figure 14, which shows nine cross sections, with x ranging from -2.5 to +2.5 at intervals of five-eighths of a meter. We can easily trace a number of features as they flow downward from one section to the next; a curve connecting similar features would represent a path taken by a sled. The complete attractor is seen to be composed of groups of nearly parallel surfaces, generally oriented more or less vertically; these are what appear as nearly parallel curves on each cross section. The continual stretching, compression, and twisting of the upper patterns to produce the lower ones is evident. The bottom cross section is of necessity the same as the top one, with the left and right halves interchanged. It is also the cross section shown in Figure 12.

An attractor that consists of an infinite number of curves, surfaces, or higher-dimensional manifolds—generalizations of surfaces to multidimensional space—often occurring in parallel sets, with a gap between any two members of the set, is called a *strange attractor.* The name was introduced in the early 1970s by David Ruelle and Floris Takens in a paper in which they proposed that fluid turbulence is an example of what we now call chaos. There have been some objections to the term, and the Russian mathematicians Boris Chirikov and Felix Izrailev have even stated that a strange attractor seems strange only to a stranger. Their point is that these infinite complexes of manifolds are precisely what anyone should have expected, notwithstanding the fact that rather few

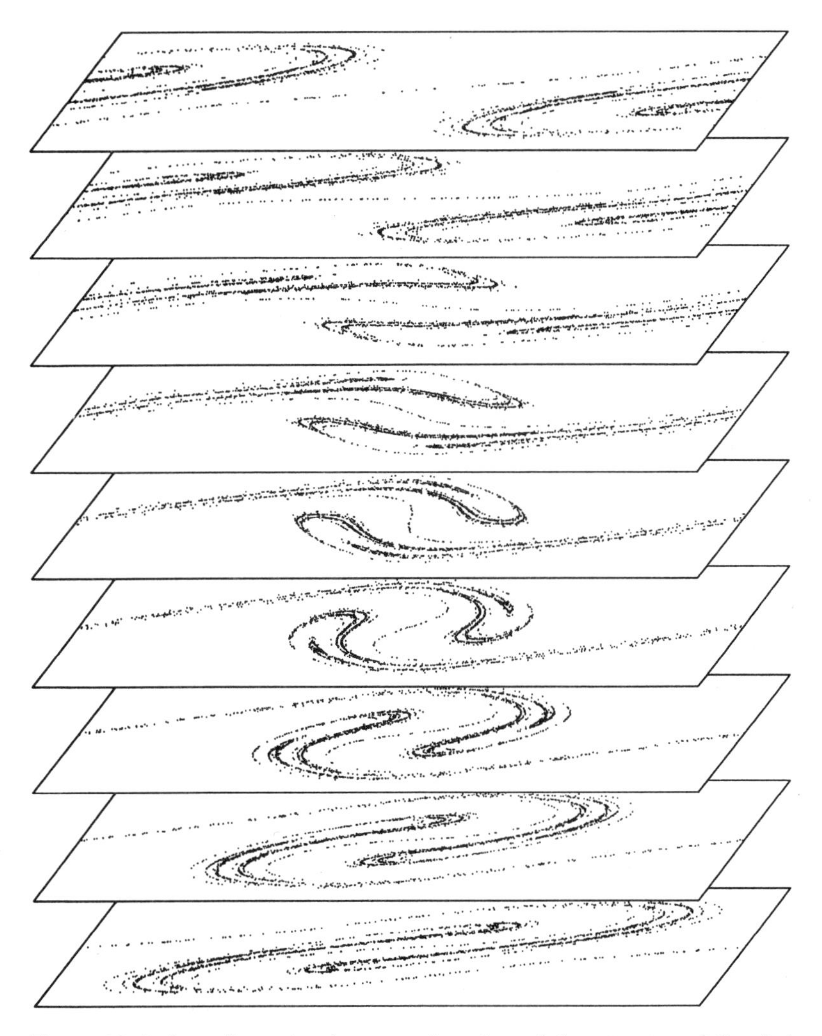

Figure 14. A three-dimensional perspective view of the attractor of the sled model, as pictured by nine parallel cross sections. The lowest section is the one appearing in Figures 12 and 13.

people did expect them. The name has nevertheless been too picturesque for most scientists to resist, and it seems to be firmly established. John Guckenheimer, one of the pioneers in the field, has even titled one of his papers "A Strange, Strange Attractor."

A strange attractor, when it exists, is truly the heart of a chaotic system. If a concrete system has been in existence for some time, states other than those extremely close to the attractor might as well not exist; they will never occur. For one special complicated chaotic system—the global weather—the attractor is simply the climate, that is, the set of weather patterns that have at least some chance of occasionally occurring.

Still, there is something counterintuitive about having possible states that are almost the same separated by impossible states, as they are in any strange attractor. It is almost like stating that the maximum temperature next Thursday, or a year from next Thursday, can be 25 or 27 degrees but cannot possibly be 26, and, if measured to the nearest tenth of a degree, can be 25.1 or 25.3, or 26.9 or 27.2, but cannot be 25.2, or 27.0 or 27.1. Such a pronouncement is unlikely to come from a weather forecaster or climatologist who is interested in continued employment.

The strange set of points in which a line can intersect a strange attractor is a simple example of a *Cantor set.* The German mathematician Georg Cantor, who pioneered the study of these and many other sets, presented the mathematical world with a famous example. Take a horizontal line segment, discard the middle third while retaining the end points, then discard the middle third of each of the two resulting segments while again retaining the end points, and continue the process to infinity. It might seem that in the limit nothing would be left but end points, but this is not the case. The point one-fourth of the way in from the left end of the original segment, for example, will, after the first step, be one-fourth of the way in from the right end of the left segment, and, after each succeeding step, will be one-fourth of the way in from one end or the other of the segment in which it lies. Thus it will never be an end point, yet it will never be discarded.

An essential feature of our model is revealed by the successive panels in Figure 11. Since each set of points is mapped by the Poincaré mapping into a portion of itself, it is *ipso facto* mapped into a set occupying a smaller area in phase space. Sets of points that occupy small areas to begin with are mapped into sets occupying even smaller areas. If an initial set of points fills a small circle, its successive maps, for a while at least, will fill a succession of approximate ellipses. Since nearby points tend to be mapped to points that are farther apart, i.e., since the system exhibits sensitive dependence, the long axes of the ellipses must eventually, if not immediately, become longer and longer. At the same time, the short axes will become shorter so rapidly that the areas enclosed by the ellipses will become smaller and smaller.

In Figure 15, which, like the pictures of the attractors, is a diagram of a cross section of phase space, the points enclosed by the circle near the upper left corner represent the initial positions and speeds of a collection of sleds; all of them are slightly east of a mogul and are moving a few degrees east of southward. The points enclosed by the curves that resemble ellipses, and that lie successively farther from the circle, represent the positions and speeds of the same sleds after they have descended 1, 2, 3, 4, and then 5 meters. The continual stretching of one axis and compression of the other is apparent. The final "ellipse," which is visibly distorted, looks almost like a segment of a curve. Its long axis has stretched about fivefold, but its short axis has been compressed more than twentyfold, so that it encloses less than a fourth of the area enclosed by the circle.

A two-variable system in which areas continue to decrease, or a more general system in which multidimensional volumes in phase space continue to decrease, whether or not there is stretching in one or perhaps several directions, is called a *dissipative* system. Tangible dissipative systems generally involve some physically damping process, like friction. Most familiar physical systems are dissipative, although some of them behave almost like systems without dissipation, and can even be advantageously studied with nondissipative mathematical models, provided that one is not interested in finding their attractors. For example, a pendulum whose motion is maintained against the damping effects of friction by the works of a clock is often treated as an unforced undamped system, and hence as if it were nondissipative.

Although the successive panels in Figure 11 are clearly converging toward a strange attractor, we may be left wondering whether this behavior is only a quirk of the system that we have been studying. A simple analysis will show that strange attractors are really rather general features of chaotic dissipative systems as long as the variables have restricted ranges. As we shall see later, a system that is not dissipative may have no attractor at all.

For definiteness, consider a two-variable system. Recall first that, in any chaotic system, two states that are almost alike will eventually evolve into states that are no more alike than two states selected at ran-

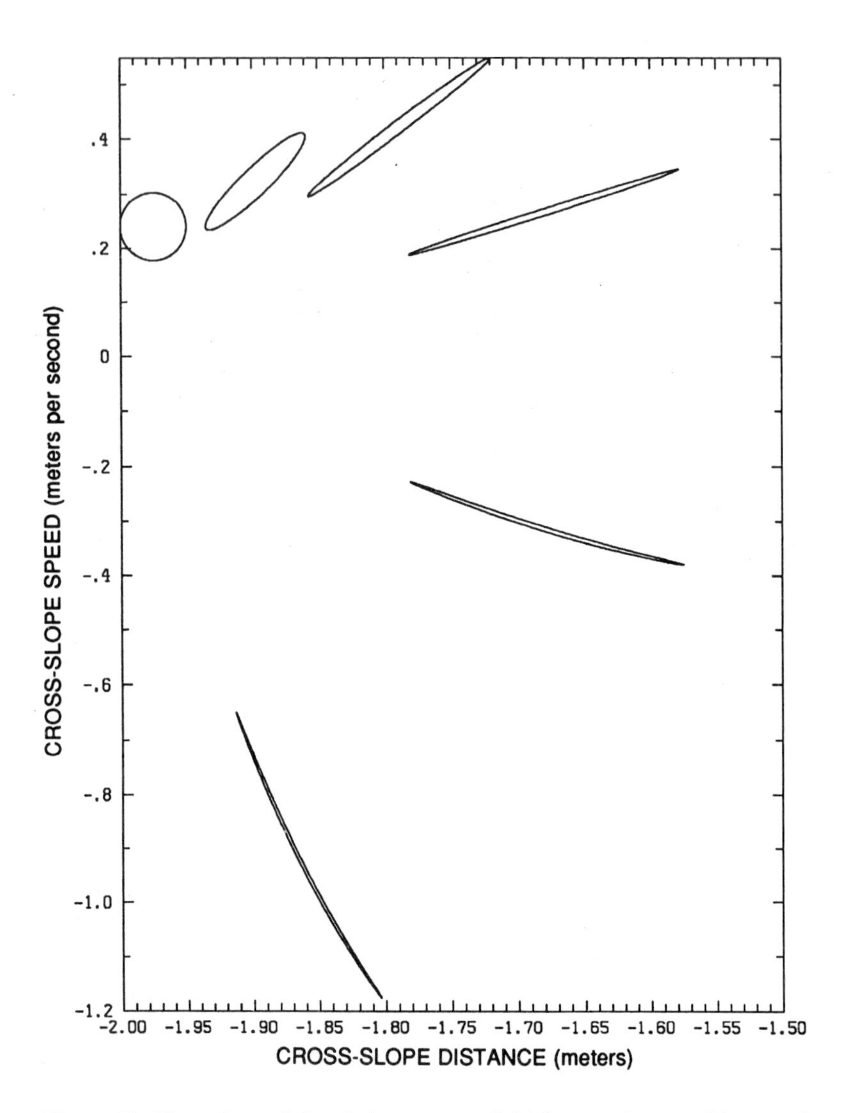

Figure 15. The points of the circle represent initial cross-slope positions and speeds of a collection of sleds. The elongated structures, in order of increasing distance from the circle, represent the positions and speeds of the same sleds at 1-meter intervals down the slope. Note that the figure covers only a small portion of the area covered by Figure 12.

dom from a long sequence. This implies that a small local region in phase space, such as the one enclosed by the circle in Figure 15, will, after some time, be deformed into a region that extends most of the way across the attractor. If the system is dissipative, the new region will have a smaller area than the original one, and, being long, will have to be narrow, and will look like a segment of a curve.

Next note that the original small region can be subdivided into many very small regions. It follows as before that each very small region will also, after enough time, be deformed into an elongated region that nearly spans the attractor, and resembles a piece of a curve. Recombining the pieces, we see that the original region will, by this time, have been deformed into a region that resembles a much longer segment of a curve. In the limit the segment will become infinitely long. Since, in being continually deformed, the region will resemble the attractor more and more closely, the attractor must resemble a curve of infinite length.

Because the variables have limited ranges, the infinitely long attractor, like the attractor pictured in Figure 12 or 13, must fit into a restricted region. It can do this most readily by doubling back on itself an infinite number of times. Its appearance will indeed be "strange."

It is evident that a straight line will intersect this attractor in infinitely many points, but it would require a more detailed investigation to show that these points must form a Cantor set—one in which gaps separate every pair of points. Pending such an investigation, one should not rule out the possibility of exceptions. Nevertheless, this brief analysis should suffice to liberate us from Chirikov's and Izrailev's company of strangers.

It is time to return to the sliding board with its four variables. Here the attractor is contained in a four-dimensional "box." A cross section of the box, say a section on which x is constant, will be three-dimensional; it will be a conventional box, and will contain the attractor of the Poincaré mapping.

In three dimensions, a strange attractor may be an infinite complex of curves, but it may also be an infinite complex of surfaces. To determine graphically which form an attractor assumes, we can draw its projection on one face of the box. In the former case we shall see curves, but in the latter the separate surfaces will project on top of each other and completely fill an area on the face.

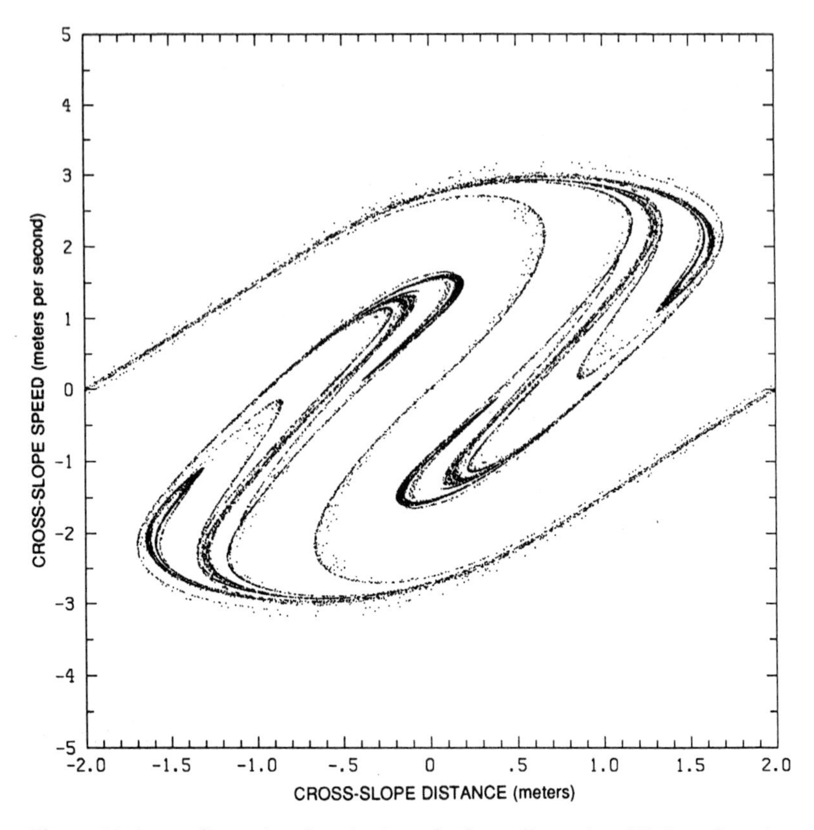

Figure 16. A two-dimensional projection of a three-dimensional Poincaré section of the attractor of the board model. The projection is obtained by plotting values of V against y and disregarding U.

In the present instance, we can find the projection of the cross section by simply repeating the computational procedure that we used in producing the attractor in Figure 12, ignoring the fact that U actually varies on the attractor. We obtain Figure 16, which bears a striking resemblance to Figure 12. We conclude that, as with the sled model, the board model has an attractor whose cross section consists of curves. If, in phase space, the direction in which U varies is assumed to be perpendicular to the page, we can produce the three-dimensional cross section by pushing certain parts of Figure 16 behind the page and pulling other parts forward. Where two curves in the figure cross, one of them may have to be pushed while the other is pulled.

The long-term average rates of stretching or shrinking of the axes of an infinitesimally small ellipsoid serve as distinguishing features of individual dynamical systems. For both the sled and the board, only the longest axis continues to stretch. In the board model, if two axes had stretched, or even if the second axis had shrunk less rapidly than the first one stretched, the attractor of the Poincaré mapping would have been composed of surfaces instead of curves. In systems with many variables, many axes can stretch. If the system is chaotic, at least one axis must stretch, but, if the system is also dissipative, some of the axes must shrink so rapidly that the volume of the ellipsoid continually diminishes.

Broken Hearts

Unlike the sled model, the board model has a second attractor. A board that starts moving very slowly near the low point of a pit, or one that is given an initial upward push and just manages to enter a pit, can become trapped, in which case it will eventually come to rest at the low point. The state of stable equilibrium that it attains is an attractor, represented in four-dimensional phase space by a single point. The model therefore does not always behave chaotically. It appears, in fact, that if a set of attractors is the heart of a system, the board model has a broken heart.

In Figure 17 we see portions of a few of the dark and light checkerboard squares that cover the ski slope. We also see a specially constructed closed curve. A board starting from rest at any point enclosed by the curve will eventually come to rest again at the low point, indicated by the central dot. Boards starting from rest at points outside the curve, but still within the realm of Figure 17, will travel down the slope. On a more extensive picture of the slope, a similar closed curve would surround each pit.

When a system has more than one attractor, the points in phase space that are attracted to a particular attractor form the *basin of attraction* for that attractor. Each basin contains its attractor, but consists mostly of points that represent transient states. Two contiguous basins of attraction will be separated by a *basin boundary.*

Since the basins in our model are fully four-dimensional, visualizing one of them may prove awkward. Taking a cross section of phase space

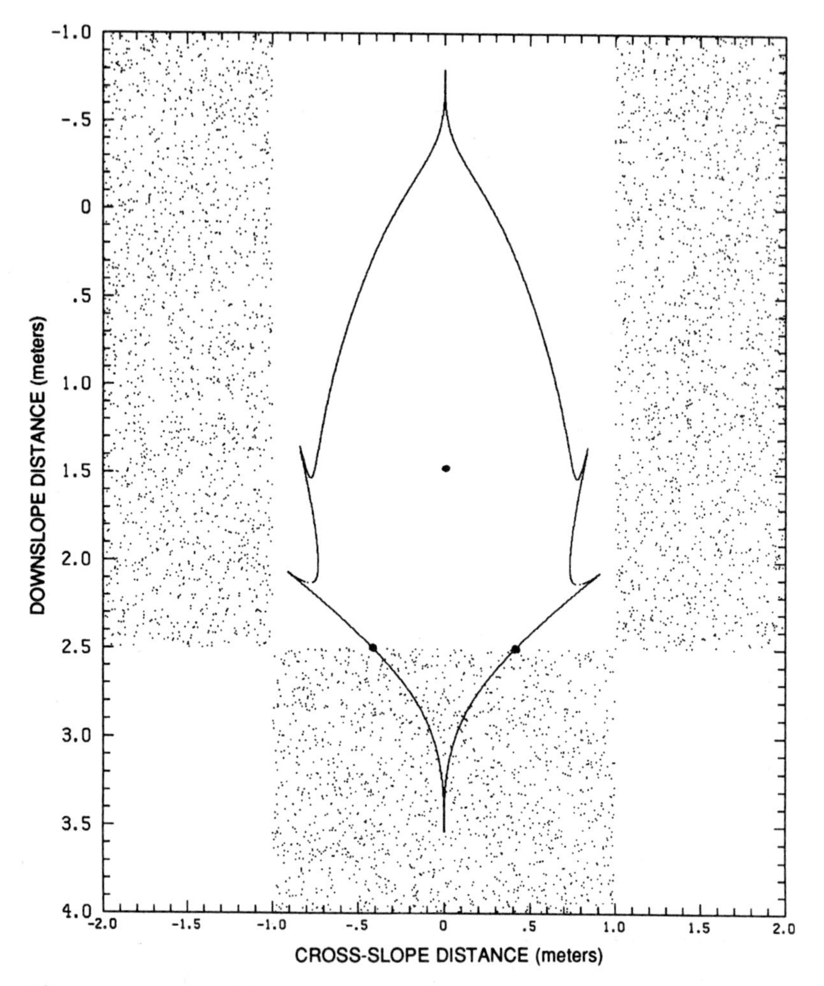

Figure 17. A section of the ski slope. The jagged curve separates the points on the slope from which a board starting from rest will become trapped in a pit from those from which it will continue down the slope. The central dot indicates the lowest point in the pit, and the other two dots indicate the saddle points.

by fixing the value of one variable, as we did in constructing Figure 16, still leaves us with three-dimensional objects, which may also be difficult to depict. We can partially solve our problem by taking a double cross section. That is, we fix the values of two of the variables, which may be any two, and we allow the other two to vary. The double cross sections of the basins will be two-dimensional, and they will be separated by a one-dimensional boundary—a curve.

For definiteness let U and V, the velocity components, be the variables whose values we fix, and let both values be zero. Let the values of X and Y equal the distances, vertically downward and horizontally to the right, from a reference point on the section. Since X and Y are also, by definition, equal to the southward and eastward distances from a pit on the slope, the double cross section of phase space must for all practical purposes be a picture of the slope. The double cross section of the basin boundary is then identical with the closed curve in Figure 17.

One might have supposed that this curve would be an ellipse or some other smooth curve, but evidently it looks more like a leaf, and possesses six distinct cusps. These may be readily accounted for.

Consider a board that starts from rest, at a point on the curve. In phase space, the point that represents its state is on a basin boundary; not being *inside* a basin, it will not approach an attractor. Instead it will do the only remaining thing possible—it will stay on the boundary. As long as the board continues to move, friction will continue to deplete its energy, and, since it could never regain this energy without continually moving down the slope, and abandoning the boundary, it must ultimately come to rest again.

Within a dark checkerboard square and the light square just to its north, there are four points at which a board can remain at rest. One of these is the low point of a pit, where the equilibrium is stable. The other three points are all on the closed curve shown in Figure 17, and the equilibria are all unstable. Only one of these points is also a cusp; it is the high point of a mogul and the southernmost point of the "leaf"—call it the south point. The other two points are located where the edge of the leaf crosses from a light square to a dark one, and are shown by dots in Figure 17. They are both saddle points—points where the slope is shaped like a mountain pass. If you move along the edge of the leaf in either direction from one of these points, you will move upward, but if you move directly across the edge, you will move downward—into the pit or down the slope.

A board placed at rest near but not exactly at the south point will begin to move in the direction of steepest descent. If it is placed due north of the south point it will slide directly over the pit and then back again, eventually coming to rest at the low point. However, because the south point lies on a ridge that is elongated in the north-south direction, so that the slope falls off more rapidly to the west and east, a board placed a short distance from the south point in any direction other than almost due north will curve around and slide off one side or the other. It will then plummet down the slope. If a board is to move neither down the slope nor into the pit—that is, if it is to be on the boundary—it must be placed almost but not exactly due north of the south point; the closer to the south point it is placed, the more nearly due north the displacement will have to be. The allowable points thus form a cusp at the south point, which is thereby accounted for.

The north point—the northernmost point of the boundary—is the point from which a board starting from rest will slide due southward, cross the pit, and just come to rest at the south point. Boards starting near the north point will therefore come nearly to rest, nearly at the south point, after which they will behave more or less as if they had been given a slight shove from the south point. Most of them will continue down the slope, but a few, starting almost due south of the north point, will be trapped, and the shape of the boundary near the north point will mimic that near the south point—hence the northern cusp.

Boards starting from rest on either side of the leaf will generally come to rest at the closer saddle point, but each side evidently has a short central segment from which a board will cross over to the opposite saddle point. These segments must be separated from the remaining portions of the sides by points from which a board will not reach either saddle point, and will therefore do the only thing that is left—come to rest at the south point. Again, the boundary close to any of these points will mimic the boundary near the south point; thus the cusps at these points are explained.

Recall now that the complete basins of attraction are four-dimensional, while the separating boundary is a three-dimensional manifold embedded in the four-dimensional phase space. States confined to this manifold effectively comprise a new, three-variable dynamical system, embedded in the larger system. Dynamical systems of this sort can possess their own attractors, as well as their own basins and basin boundaries if they have more than one attractor. In the present case, the new system has two attractors—states of rest at the saddle points. Their basins are three-dimensional manifolds whose double cross sections are the smooth segments of the sides of the leaf, while their basin boundary is a two-dimensional manifold whose double cross section consists of the tips of the six cusps. In a simple rather than a double cross section the cusps would appear as knife edges. To see their complete structure would require four-dimensional vision.

Since the new system has its own basin boundary, states confined to this boundary form a still smaller dynamical system, embedded in the embedded system. It has a single attractor—a state of rest at the south point—and here the embedding process ends.

Basin boundaries of more general dynamical systems can have very complicated structures, and even in the present example they may be more complicated than they appear to be. As boundaries or as new dynamical systems, they constitute another aspect of dynamical-systems theory that has appealed to many scientists.

Systems can have multiple attractors without being chaotic. In both the board model and the sled model, if the height of the moguls above the neighboring pits is considerably reduced, there will be two attractors, each corresponding to a periodically wiggling path, one progressing southeastward and one southwestward. Each attractor will consist of a single curve, and each will appear as a single point in a Poincaré cross-section.

Attractors are examples of *invariant* sets—sets that will consist of precisely the same points if each point is replaced by the point to which it is mapped. When there is more than one attractor, each basin of attraction is an invariant set, as is the basin boundary, sometimes called a *separatrix*. There is still another invariant set, which connects the attractors when there are more than one, and which by analogy ought to be called a "connectrix," but generally, together with the attractors that it connects, is called the *attracting set*. Despite its name, the attracting set should not be confused with the set of attractors, which is sometimes only a portion of it. To construct an attracting set we can proceed as we did in constructing Figure 11; we start with a large collection of points, supposedly representing an infinite collection, and determine what happens when these points are continually replaced by their images. The limiting set will be the attracting set.

The attractor being approached in Figure 11 is therefore the attracting set also. Figure 18 has been constructed similarly, except that the moguls rise 50 instead of 100 centimeters above the pits, and the sleds have been allowed to descend 10 meters instead of 5 between successive panels. By

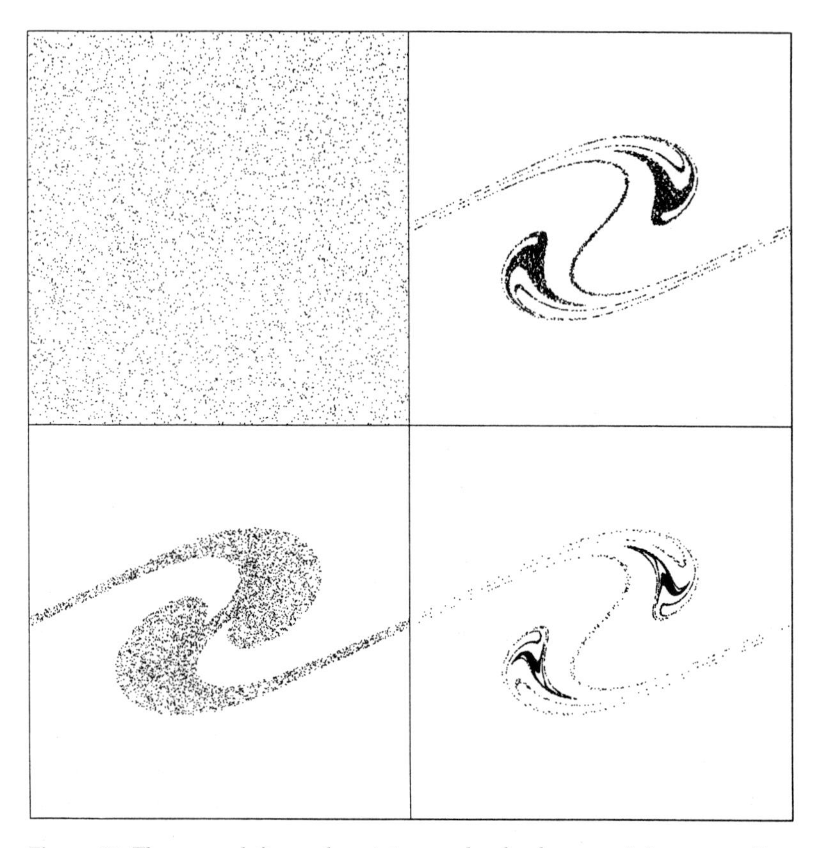

Figure 18. The upper left panel contains randomly chosen points representing the cross-slope positions and speeds of five-thousand sleds. The height of a mogul above a pit to its west or east has been reduced to 50 centimeters. The lower left and then the right-hand panels represent the positions of the same sleds after they have traveled 10 and then 20 and 30 meters down the slope.

30 meters, most of the points have moved to one or the other of two small dark patches. The original five thousand points were selected randomly, and it is probable that if the mapping were to be continued indefinitely, every one of them would end up at one or the other of two points—the attractors—one in each dark patch. However, if we could have started with *every* point in the rectangle, there would at each iteration have been some points remaining on the infinitely long, infinitely twisted thread—the remainder of the attracting set—that connects the two attractors.

Chaos of Another Species

Let us take another look at that paradigm for well-behaved dynamical systems—the pendulum in the clock. Friction is removing energy, quite slowly but quite surely, and the clockwork is replacing the lost energy, perhaps by giving the pendulum a slight tug at the extremity of each swing. If we plot the speed of the bob against its displacement from the vertical, we obtain a simple closed loop resembling an ellipse—a phase-space representation of all of the states that the pendulum assumes during its swing.

As we saw earlier, the loop also represents the attractor. If we give the pendulum a push, it will respond with a wider swing, but within perhaps a minute it will be swinging as before; the new states will have been attracted to the original loop.

Now imagine that, while the pendulum is swinging normally, we can somehow turn off the force of friction. Let us simultaneously remove the clockwork, which will no longer be needed. The pendulum will then continue to swing almost as if nothing had happened, and its state will trace out nearly the same loop. It is therefore not surprising that an unforced, undamped pendulum has often served as a model of what takes place in a well-behaved clock. Nevertheless, as dynamical systems the model and the real-world pendulum do not have much in common. For the frictionless pendulum, the closed loop is *not* an attractor; states that are not already on the loop are never attracted toward it. If we give the pendulum a push, the resulting wider swings will persist, and the loop will be replaced by a larger loop, until such time as we disturb the pendulum again. It will be useless to wait for transient effects to die out; no states are transient. The collection of points in phase space occupying the area between any two concentric loops will continue to occupy this area, and hence will never be squeezed into a smaller area.

Systems in which volumes in phase space, or areas if the system has only two variables, neither decrease nor increase as time progresses are called *volume-preserving*. Systems in which some quantity, such as total energy, retains a fixed value as time advances are called *conservative*. Conservatism and volume preservation frequently go hand in hand, and systems possessing both properties are often called *Hamiltonian*, although the systems that conform to the equations formulated by the Irish mathematician William Rowan Hamilton are somewhat more specialized. Models of familiar real physical systems that simply disregard all dissipative processes and all energy sources are generally Hamiltonian. Probably the most familiar real-world, or real-universe, Hamiltonian system consists of the sun with its planets orbiting about it. As with the frictionless pendulum, widespread sets of points in phase space will not converge upon smaller sets, and there will be no attractors.

Hamiltonian systems may be chaotic; note that the qualitative reasoning indicating that a pinball machine should behave chaotically does not invoke any dissipation. Chaotic systems therefore do not always possess strange attractors, although most of the generally encountered compact dissipative chaotic systems do have them. Despite the absence of these intriguing features, many scientists have chosen Hamiltonian systems as the ones that they prefer to study.

To discover why this should be so, let us convert the board on the ski slope into a Hamiltonian system. We can do this, mathematically, by removing the friction, and also removing the general southward drop in elevation, which plays a similar role to the clockwork that drives the pendulum. The moguls and pits will then project from a horizontal surface. We may even suppose that we have changed the old system to the new one suddenly, after the board has followed its path long enough to shed any transient effects. We may then be tempted to conclude that, if we make no further changes, our action will have an inconsequential effect on the ensuing motion, just as with the pendulum, but a visit to the computer will show us that this is not the case at all.

The energy of a board consists of kinetic energy, represented by its speed, and potential energy, represented by its elevation above the bottom of a pit, and it is the sum of these two forms of energy that remains fixed as time advances. The total energy is therefore a constant of the model, and one could logically think of the system as consisting of a family of dynamical systems—one system for each value of the energy. Let us consider in detail the possible behavior of a collection of boards, all having the same total energy, and hence, in the present sense, all belonging to the same system. If the boards have very little energy, they will be trapped in light-colored squares of the checkerboard—the ones displayed in Figure 6—that surround the pits. If they have somewhat more energy, they can escape through corners to other light squares, but cannot penetrate far into the dark squares. If they have still more energy, they can go anywhere. For definiteness let us examine an intermediate

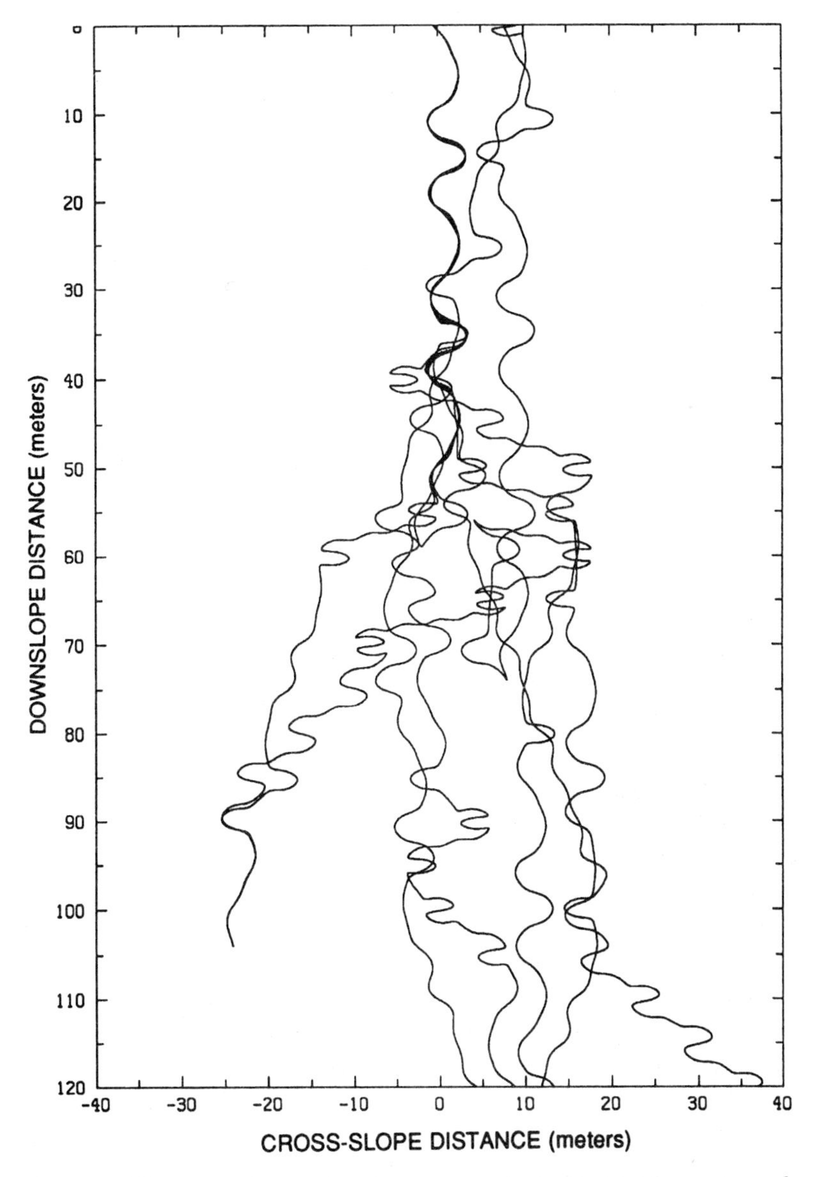

Figure 19. The paths of seven boards starting with identical total energies and identical velocities from points spaced at 1-centimeter intervals, on the Hamiltonian ski "slope" with no friction and no continual southward drop.

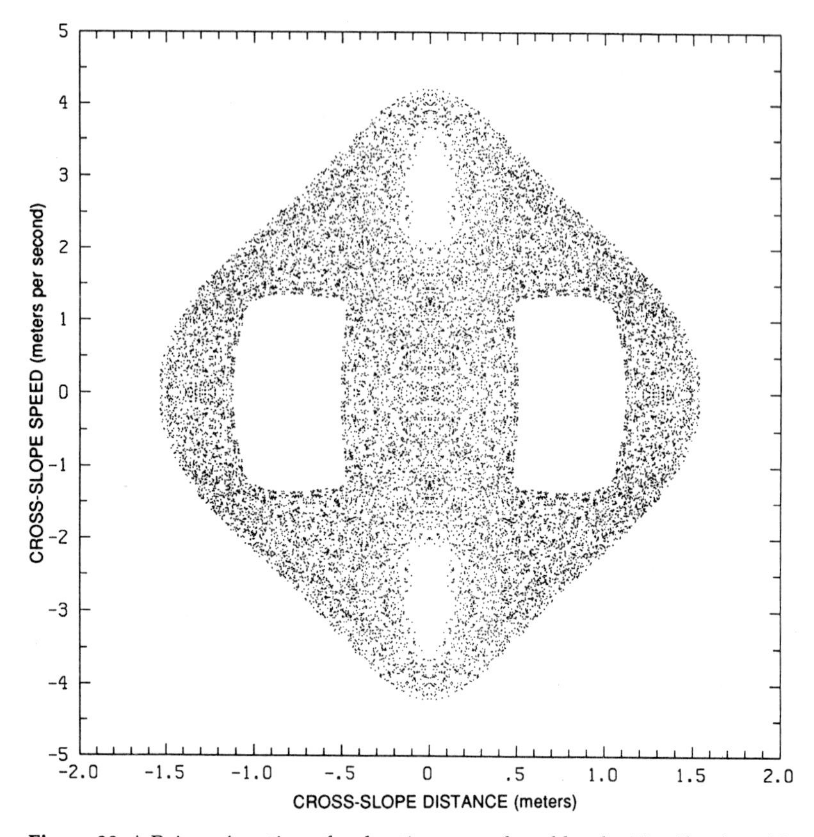

Figure 20. A Poincaré section of a chaotic sea produced by the Hamiltonian ski-slope model, when the total energy is nine-tenths of that needed to reach the top of a mogul.

case, in which the energy is nine-tenths of the minimum amount needed to reach the top of a mogul.

In working with the dissipative ski-slope model, we first detected chaos by looking at the paths of seven boards that started with rather similar conditions, and noting that they rapidly diverged. We may study the new model similarly. Figure 19 shows what can happen; here the starting points are 1 centimeter apart, and appear in the figure to be all the same. Before 50 meters the paths have visibly diverged, confirming the presence of chaos, but thereafter they proceed so erratically that when they are displayed together it is not easy to see which ending follows from which beginning. Two of the boards even turn around and

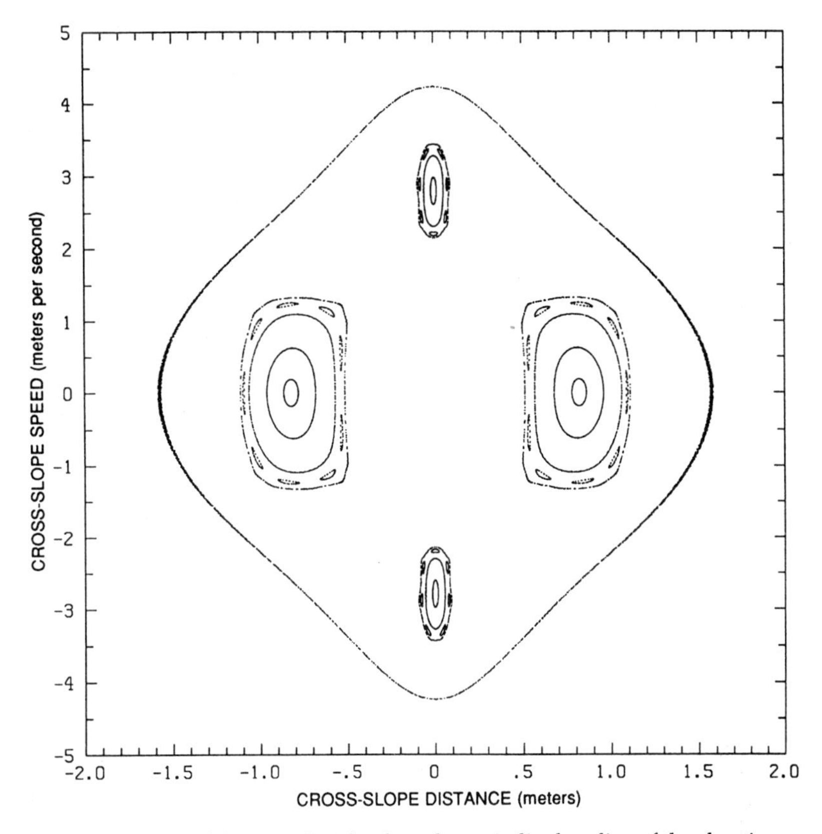

Figure 21. Some of the periodic islands and a periodic shoreline of the chaotic sea of Figure 20.

slide back across the starting line—something that could not possibly happen in the dissipative model. Even though the new system is conservative, in that it preserves the total energy of each board, there is nothing in it to preserve the original southward progression, as there was in the dissipative model.

There are various ways to display the long-term properties of these and other paths, but it is particularly simple to do just as we did with the dissipative model. We choose an initial point, and then plot the cross-slope speed against the cross-slope position , this time whenever the board crosses a west-east line through a pit and a mogul, i.e., whenever x equals 0. For the dissipative system the procedure produced a cross section of a strange attractor; we have already seen that for any Hamiltonian system it must do something else, since there will not be an attractor.

Figure 20 shows what happens when the initial point is the start of one of the paths in Figure 19. Not surprisingly, the points appear to fill an area, sometimes called a *chaotic sea,* instead of lying on separate curves with gaps between them. What may surprise us is the four prominent holes. It seems rather unlikely that such large areas would be missed if they were to be eventually occupied.

If it is true that the sequences of points—representations of states at a succession of key locations along a path on the slope—will never enter a hole, it must be equally true that other sequences beginning in a hole will never enter the sea. This follows, because a possible *path* remains a possible path if the direction of motion along it is reversed—note what happens near the southwest corner of Figure 19—so that a possible *sequence* of points remains a possible sequence if the order is reversed. Let us therefore repeat our procedure a number of times, choosing in each instance an initial state in one of the holes.

We obtain the composite picture in Figure 21, containing four patches that would fit into the holes in Figure 20. Each new sequence produces a closed loop, or else a chain of small loops surrounding a larger loop. The points of a sequence do not progress continuously along a loop; they jump from one location to another, generally in jumps of more or less equal length, and, having completed a circuit, they proceed to fill in the gaps.

Figure 22 is a blow-up of the uppermost patch, with additional horizontal stretching. The loops shown have been produced by six sequences of points. One sequence produces the chain of seven loops that was barely visible before, while another produces the seven smaller loops inside them. Outside the outermost large loop shown is another chain of small loops—this time nineteen of them. They have been produced by a single sequence of points, which visits all nineteen loops before returning to the first. The boundary of the chaotic sea, not shown, lies very shortly beyond these loops, but before it is reached the pattern that we have seen—more large loops surrounding the chains of small loops and more chains of small loops surrounding the large loops—repeats itself an infinite number of times. There are also chains of very small loops surrounding each small loop. This sort of structure, which

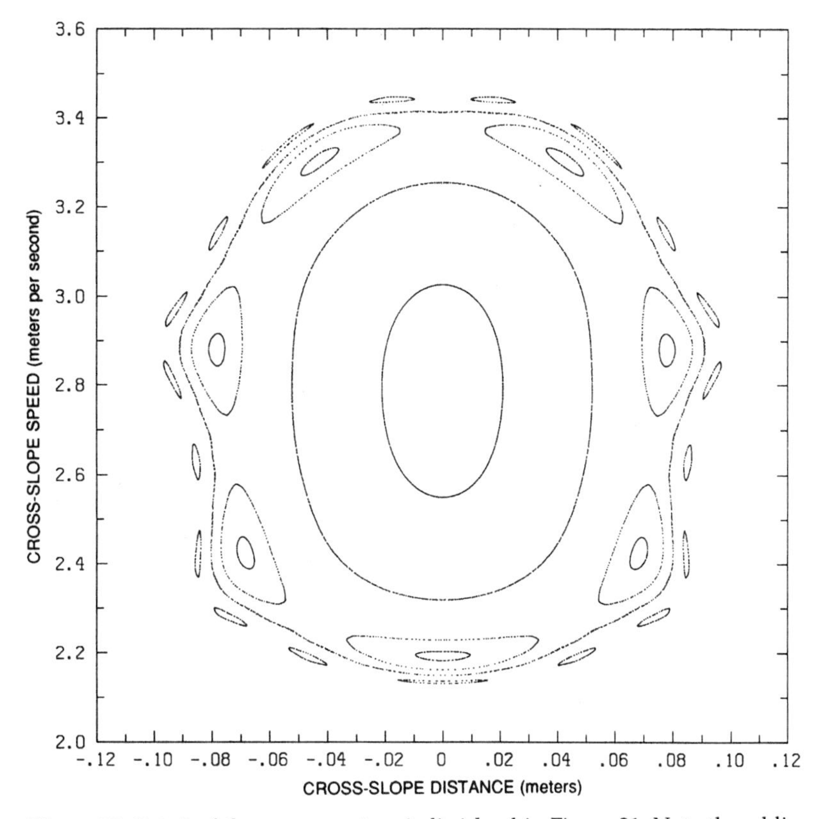

Figure 22. Detail of the uppermost periodic island in Figure 21. Note the additional horizontal stretching.

can be as fascinating to study as that of a strange attractor, appears rather generally in Hamiltonian systems, with individual variations. Note that in the larger patches in Figure 21 the first chains consist of nine loops instead of seven.

Suitably chosen initial states will produce single points at the centers of concentric loops. These represent periodic paths, while the loops themselves represent almost-periodic paths. The large loop in Figure 21 that would surround the sea was not added for aesthetic purposes; it corresponds to another almost-periodic path, along which a board will shuttle across a pit between the sides of two moguls. Dissipative systems are often entirely chaotic, with strange attractors, or entirely periodic, like the pendulum with its simple attractor. In Hamiltonian sys-

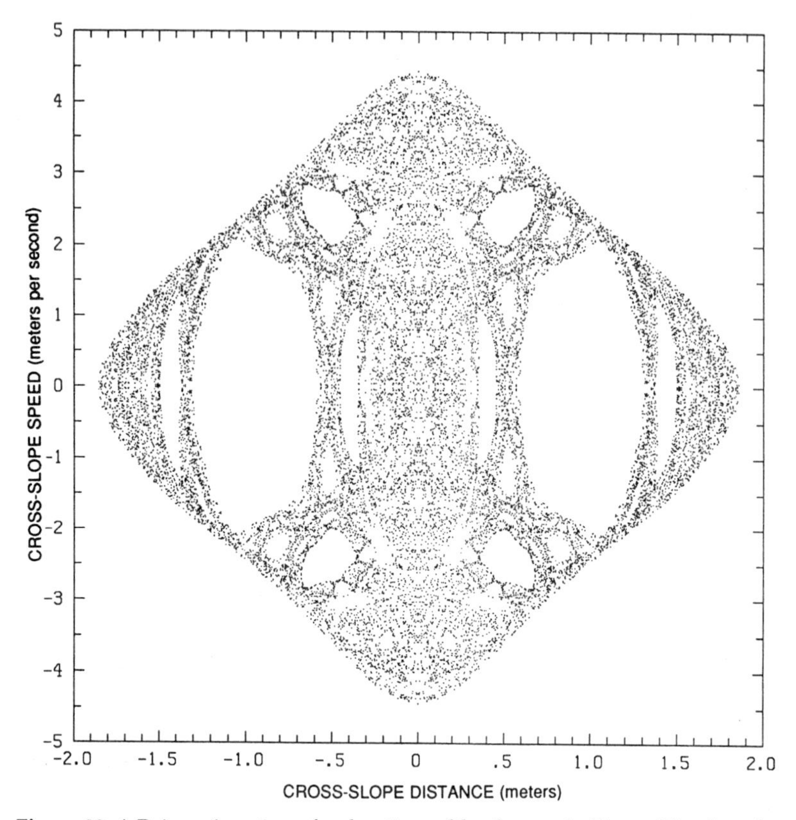

Figure 23. A Poincaré section of a chaotic sea like the one in Figure 20, when the total energy is 99 percent of that needed to reach the top of a mogul.

tems, even for a single value of total energy, it is common for some initial states to lead to almost-periodic variations while others lead to chaos.

For other choices of total energy, periodic paths like those corresponding to centers of loops can be unstable, in which case initial states chosen close to them will lead to chaotic paths, and prominent holes like those in Figure 20 will not appear. Still other choices of total energy can produce still more holes; this is especially likely when the energy is just about sufficient for a board to reach the top of a mogul, and the gentle slope that it encounters when it is hardly moving tends to favor regularity. Figure 23 is constructed in the same manner as Figure 20, but now the total energy is 99 percent of that needed to cross a mogul. The detectable holes are far more abundant, and the big ones have become bigger, although the upper and lower ones seem to have disappeared.

In and Out of Chaos

At a symposium that I attended some time ago, when chaos was already drawing considerable attention but its ubiquity was yet to be recognized, a session chair invited the audience to suggest some timely topics requiring additional study. Someone volunteered "routes to chaos." I found myself in disagreement, not with the topic, which was undeniably important, but with what it had been called. I sensed an implication that, in order to account for chaos, we would need to know how it could arise out of some more "normal" behavior—regular behavior—if some feature of a family of dynamical systems were to be progressively altered. I felt that it was inappropriate to regard regular behavior as the more fundamental type, and that it would be just as logical to maintain that in order to understand regularity we would need to know how it could evolve from chaos. I suggested that we might equally well call the topic "routes from chaos."

It is true that in many families of forced dissipative dynamical systems which possess only states of rest when the forcing is absent, regular behavior will be the first to develop when the forcing is raised step by step to higher values, while chaos will not appear until later, if it appears at all. Yet in some real systems, such as the global circulation of the atmosphere, the forcing is and always has been strong; there is no reason to believe that the weather once behaved like clockwork before it became chaotic, and there is no need to postulate the existence of a route from some form of regularity to the chaos that now prevails.

Nevertheless, there is no question but that the chains of events through which chaos can develop out of regularity, or regularity out of chaos, are essential aspects of families of dynamical systems. A common response to a slight alteration in some constant is a slight modification of the attractor and slight changes in other properties. Sometimes, however, a nearly imperceptible change in a constant will produce a qualitative change in the system's behavior: from steady to periodic, from steady or periodic to almost periodic, or from steady, periodic, or almost periodic to chaotic. Even chaos can change abruptly to more complicated chaos, and, of course, each of these changes can proceed in the opposite direction. Such changes are called *bifurcations.*

Bifurcations can occur in various ways. A state of stable equilibrium may be rendered unstable when some constant is increased. If originally the state was an attractor, it will cease to be one; small disturbances will amplify and produce a new mode of behavior. Changes from instability to stability are equally possible. A hypothetical frictionless top—a Hamiltonian system, incidentally—that is spun slowly will be in equilibrium if it is standing vertically, but it will nevertheless soon fall over because the equilibrium is unstable, while a top that is spun rapidly will continue to stand, and will merely wobble if it is slightly disturbed.

Alternatively, a mode of behavior may cease to exist altogether. A board that has come to rest at the low point of a pit on the ski slope illustrates this type of bifurcation. If the height of the moguls above the neighboring pits—call it h—is reduced while the other constants remain the same, the high points on the moguls and the low points in the pits will disappear, evidently when h falls just below 80 centimeters, and there will no longer be positions of equilibrium, stable or unstable. A bifurcation where a mode of behavior suddenly goes out of existence, rather than simply becoming unstable, is called a *saddle-node* bifurcation.

To examine sequences of bifurcations that can occur when a constant is varied over a considerable range, let us turn again to the board on the slope, this time when it is moving too rapidly to become trapped in a pit, and see how its behavior changes as h varies. Figures 24 and 25 are bifurcation diagrams. The vertical coordinate is h, expressed in centimeters, while the horizontal coordinate shows the temporary maxima of V, that is, the highest eastward or lowest westward speeds that the board acquires on its successive oscillations—successive excursions from a low value to a high value and back to a low one. Note that the constant being altered is not the driving force; it simply modifies a downslope progression that would occur in any case.

Where the figures show smooth curves the behavior is periodic. Each oscillation is just like the last one, or some previous one, so that only one or only a few distinct maxima of V can correspond to any chosen value of h. Where a region appears shaded, such as near the top of either figure, there is chaos, and on successive oscillations V can reach a peak at virtually any value within an extensive range.

Figure 24 was constructed by varying h in very small increments from zero, when the moguls and pits are absent, to 120, when they are a bit larger than in our original example. Figure 25 was produced similarly, except that h was successively decreased from 120 to zero. A glance at either figure reveals extensive ranges of h within which the behavior of the board undergoes no qualitative and often only minor quantitative changes, but there are a few values at which the behavior changes abruptly; these are the bifurcation points. By and large the two figures are the same, but there is a notable discrepancy in the range from 22 to 44 centimeters. This and other features of the figures are readily accounted for, once observed, although numerical computations are needed to determine where, and in some cases whether, they will occur.

Turn first to the very center of Figure 24. The curves on either side that are indistinguishable from slightly tilted straight lines, and are also present in Figure 25, represent different but symmetrically related periodic paths; on one of them the board wiggles southwestward between the moguls while on the other it wobbles southeastward. When h reaches 73, the periodic behavior becomes unstable. The board will overshoot, only to undershoot on the next wobble, until a new stable pattern of behavior becomes established, with alternate weak and strong oscillations. Even though the oscillations may retain their original period, the interval between repetitions will have doubled. The system has undergone a *period-doubling* bifurcation.

As h is increased still more, the new form of oscillation becomes unstable, and again the period doubles; this is apparent when h reaches 88. Although the present figure lacks sufficient resolution to show what happens next, the period actually continues to double infinitely many times, after successively smaller increments of h, until, at the culmination point, chaos sets in.

Sequences of period-doubling bifurcations, ending in chaos, are ubiquitous features of dynamical systems. They are not confined to mathematical models, and have been accidentally encountered or intentionally sought in a wide variety of laboratory experiments. A would-be equestrian on a trotting horse, who has just learned to post, rising up from the saddle on one step and settling back on the next, instead of bouncing up and down on each step, will be thankful for period doubling. In the present instance, but not in general, the chaos that supersedes the period doubling soon gives way to another form of periodic behavior, at 91 centimeters.

Next, turn to the base of Figure 24, where the only indicated maximum value of V, the cross-slope speed, is zero; this implies that the board is sliding due southward, along a line passing through the centers of the moguls and the pits. Near a mogul the board is riding a ridge, and,

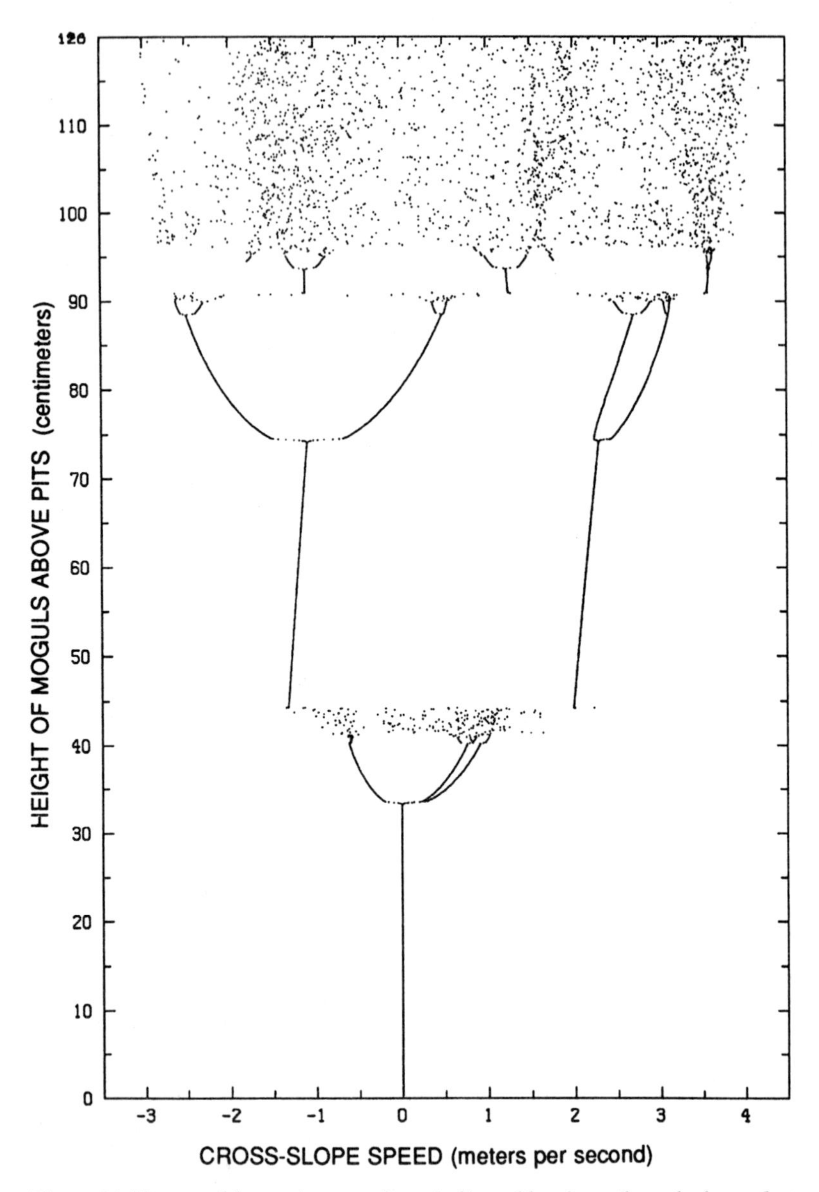

Figure 24. The possible maximum values, indicated by the scale at the base, that V, the eastward speed of a board, can assume on individual oscillations as the board slides down the slope, when the height h of the moguls above the adjacent pits assumes the value indicated by the scale at the left. The values of V have been found by increasing h in small steps.

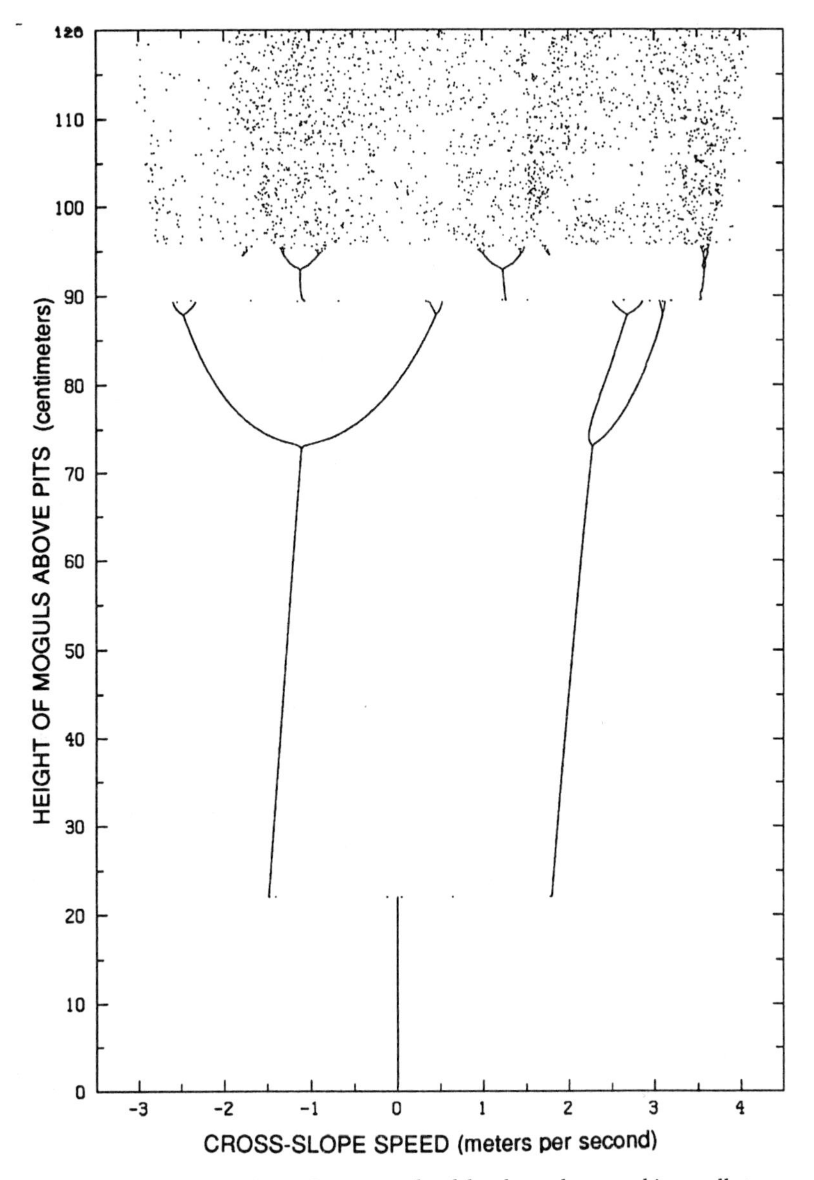

Figure 25. The same as Figure 24, except that h has been decreased in small steps.

if it becomes displaced slightly westward or eastward, it will tend to slide off. Before the board can slide very far it will be nearing the next pit, so that it will slide back again, and, aided by the damping effect of friction, it will regain its straight path. However, it will necessarily move more slowly over the moguls than through the pits, and will therefore spend more time near the moguls than near the pits, and the difference in speeds will become greater as the moguls are made higher and the pits are made deeper. When h reaches 33, the board will stay near the moguls too long for their destabilizing effect to be offset by the stabilizing effect of friction, and the straight-line path will become unstable. The board must do something new, and evidently, as seen in the figure, it will undergo a three-phase oscillation, with no net progression toward the west or east.

It appears that this oscillation also undergoes period doubling, barely detectable in the figure, and then chaos sets in. The board continues to oscillate irregularly back and forth across the north-south line, but when h reaches 44, it will swing so wide on one oscillation that it will fail to return, becoming trapped in a southwestward or southeastward route between the moguls. Chaotic behavior will no longer exist, except as transient behavior; it will have bifurcated to periodicity.

Turning to Figure 25, we find that the periodic oscillation at 44 centimeters is stable, and persists as h is lowered. Not until h reaches 22 do the oscillations disappear, not because they become unstable, but because the board can no longer move rapidly enough eastward or westward when leaving a pit to catch the next pit to the southeast or southwest; i.e., there is a saddle-node bifurcation. There is thus a considerable range of h for which the corresponding dynamical system has two possible modes of behavior, with two attractors—three, if we distinguish between southwestward and southeastward progression—each with its own basin of attraction.

Returning to 92 centimeters, we see that the three-phase oscillation is stable there also, so that, if h is decreased, the chaos that appeared when h was increased will not redevelop. Again, in a narrow range of h, the dynamical system has distinct attractors with distinct basins. If instead h is increased, period doubling occurs again, and the chaos that soon sets in tends to persist.

Actually even this chaotic range is filled with periodic "windows." They are not well resolved by the figures, and the narrower ones were in

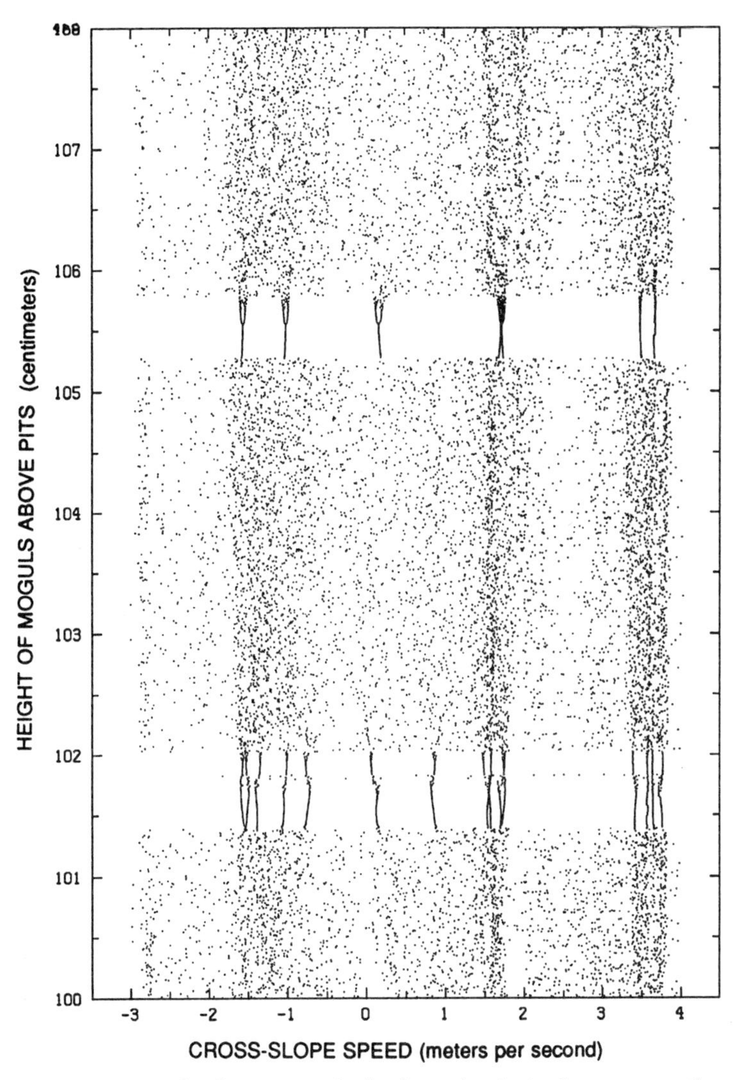

Figure 26. A vertical enlargement of a horizontal strip in the upper portion of Figure 24, revealing two periodic windows not originally resolved.

all likelihood jumped over as h was increased or decreased in steps, but two of them are quite evident in Figure 26, produced by increasing h in smaller steps, and stretching the vertical scale fifteenfold. Period doubling is detectable in the vertical segments that traverse the upper window, between 105 and 106 centimeters. The segment farthest to the left

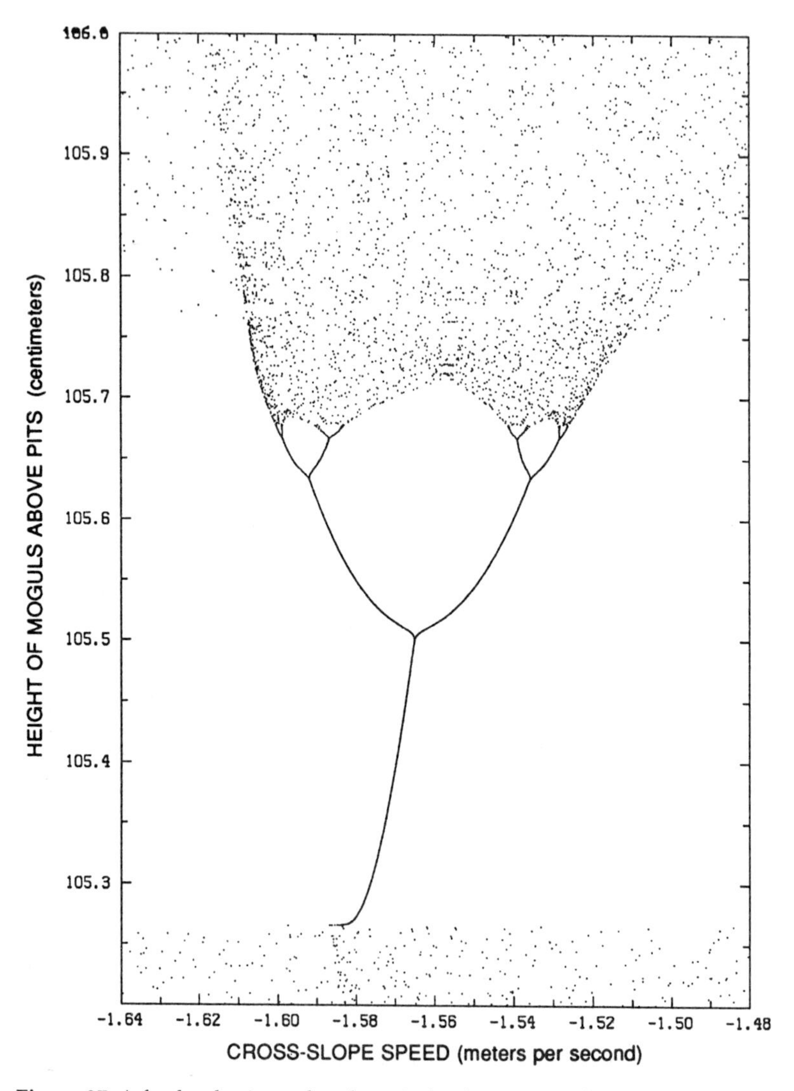

Figure 27. A further horizontal and vertical enlargement of a narrow section of the upper periodic window in Figure 26, revealing three successive doublings in a period-doubling bifurcation sequence.

has been stretched an additional ten times vertically and fifty times horizontally in Figure 27, and three successive doublings are resolved. Passing in and out of chaos again and again is yet another ubiquitous feature of families of dynamical systems.

CHAPTER 3
Our Chaotic Weather

Prediction: A Tale of Two Fluids

AN OLD JOKE about the supposedly inept weather forecaster never ceases making the rounds in one form or another. I recall one version that appeared shortly after Harry Truman, against all predictions except his own, had defeated Thomas Dewey for the presidency of the United States in 1948. It was a cartoon showing an applicant at a Weather Bureau employment office, and the interviewer was saying to him, "You worked for a public-opinion poll? I think we could use you."

Joking aside, however, the weather-forecasting community and the general public are both acutely aware that official forecasts, including those for later in the same day, are sometimes just plain wrong. To the often-heard question, "Why can't we make better weather forecasts?" I have been tempted to reply, "Well, why should we be able to make any forecasts at all?"

Why indeed should we expect to see the future, or at least a little part of it? First of all, we may believe that there is a set of physical laws governing the changes in the weather from one moment to the next, and that the very existence of these laws, whether or not we know them, ought to make prediction possible. Our faith may be fortified by the realization that other natural phenomena, governed by somewhat similar laws, have been regularly predicted with considerable success; consider the tides in the ocean, which we can predict rather accurately a few days ahead and almost as accurately many years ahead. Finally, our weather forecasts *are* correct far more often than they would be if they were pure guesses.

I must admit that my first encounter with the question of tidal prediction left me with an uneasy feeling. Apparently I had always looked at announcements of the times of coming high and low tides as statements of fact, as firmly established as the times of yesterday's tides, and not to be doubted. Yet it is apparent that they are predictions, and so, for that matter, are such "facts" as the times of sunrise and sunset that appear in almanacs. That the latter times are simply predictions with a high probability of being almost exactly right, rather than facts, becomes evident when we realize that an unforeseen cosmic catastrophe—perhaps a collision with an asteroid—could render them completely wrong. Even without a catastrophe they can be slightly in error; increases or decreases in the strength of the globe-encircling westerly-wind currents, which occur at irregular intervals, are compensated by small but measurable decreases or increases in the speed of rotation of the underlying earth, and the times of coming sunrises and sunsets may be delayed or advanced by a millisecond or so.

Returning to the ocean, let us for purposes of comparison with the atmosphere define the height of the tide as the height of the ocean surface above some fixed reference level, after individual waves have been averaged out. Let us compare the predictability of the tides, so defined, with the predictability of the temperature of the air—possibly the weather element whose prediction interests us most, although sometimes we may be more concerned with whether or not it is going to rain.

Both the atmosphere and the ocean are large fluid masses, and each envelops all or most of the earth. They obey rather similar sets of physical laws. They both possess fields of motion that tend to be damped or attenuated by internal processes, and both fields of motion are driven, at least indirectly, by periodically varying external influences. In short, each is a very complicated forced dissipative dynamical system. Perhaps it would be more appropriate to call them two components of a larger dynamical system, since each exerts a considerable influence on the other at the surface where they come into contact. The winds, which vary with the state of the system, produce most of the ocean's waves, and help to drive the great currents like the Gulf Stream. Evaporation from the ocean, which also varies with the state of the system, supplies the atmosphere with most of the moisture that subsequently condenses and still later falls as rain or snow. Why, with so many similarities, should we have had so much more success in tidal than in weather prediction? Are oceanographers more capable than meteorologists? As a meteorologist whose close friends include a number of oceanographers, I would dispute any such hypothesis.

Take a look at the periodic external driving forces—primarily the heat emitted by the sun and the gravitational pull of the sun and the moon. The atmosphere and the ocean will respond to these forces by undergoing periodic oscillations, but, as with many dynamical systems, these oscillations will be accompanied by additional irregular behavior. Near the coast, the regular response of the ocean includes most of the tidal oscillations, while the irregular response includes the occasional anomalously high tides produced by unanticipated strong winds. The regular response of the atmosphere includes the normal excursions of temperature between summer and winter, or day and night, while the irregular response includes extended hot spells and cold spells, as well as the sudden temperature changes that often accompany the progression of large storms across the oceans and continents.

It appears, then, that in attempting to forecast the tides we are for the most part trying to predict the highly predictable regular response. We may also wish to predict the smaller irregular response, but even when we fail to do so we have usually made a fairly good forecast. In forecasting the weather, or, for definiteness, the temperature, we usually take the attitude that the regular response is already known—we know in advance that summer will be warmer than winter—and we regard our problem as that of predicting those things that we do not already know simply by virtue of knowing the climate. In short, when we compare tidal forecasting and weather forecasting, we are comparing prediction of predictable regularities and some lesser irregularities with prediction of irregularities alone. If oceanographers are smarter than meteorologists, it is in knowing enough to pick a readily solvable problem. I should hasten to add that most oceanographers are not tidal forecasters anyway, nor, for that matter, are most meteorologists weather forecasters. In most respects the oceans present just as many challenges as the atmosphere.

In the following pages I shall introduce the atmosphere as an example of an intricate dynamical system, and present the case for believing that its irregularities are manifestations of chaos. After a brief overview I shall enumerate various procedures through which the presence of chaos might be confirmed. Finally, I shall examine some of the consequences of the atmosphere's chaotic behavior.

Meteorology: Two Tales of One Fluid

Chaotic dynamical systems come in many sizes. Our mathematical model of the sled on the ski slope has only three variables, yet it serves to illustrate many of the basic properties of chaos. It is by no means the simplest system with this capability; that honor, together with the honor of being the most intensively studied of all chaotic dynamical systems, goes to one with only one variable, which varies in accordance with a single quadratic difference equation—the kind of equation that defines a mapping.

The system is so simple that with a pocket calculator you can convince yourself in a few minutes that it can behave chaotically. Choose a fixed number; call it c. Choose a number between $-c$ and c as the leading member of a sequence, and construct the remainder of the sequence by always squaring the most recent number and then subtracting c to obtain the next number. Some but not all choices of c between 1.4 and 2.0 will furnish you with sequences that lack periodicity. With one of these choices for c you can also observe sensitive dependence directly, by repeating your calculations with a slightly different initial number. In a slightly altered form this system is known as the *logistic* equation, and it has been used in studies of population dynamics.

At the other extreme are systems with a large or even infinite number of variables. Among these we may expect to find the one whose states are simply the global weather patterns. Let us call it the *global weather system*.

Meteorologists have kept pace fairly well with their contemporaries in the art of creating esoteric terminology to describe esoteric concepts. They talk freely about potential pseudo-equivalent temperature, moist semigeostrophy, and dynamic anticyclogenesis, and they have even devised triple acronyms: GOCC stands for GATE Operations Control Center, GATE in turn stands for GARP Atlantic Tropical Experiment, and GARP is the Global Atmospheric Research Program, a multinational effort conceived in the 1960s and flourishing in the seventies and eighties. Nevertheless, the variables of the global weather system include no mysterious quantities. They are the familiar weather elements that we have always known—the ones that make us acutely aware of their presence, and whose values we can often estimate with fair accuracy, whenever we step outdoors. They are the temperature, the wind, the humidity, and some representation of the clouds that may be enveloping us and the rain or snow that may be falling on us. To these we must add pressure—a familiar item in many weather reports even though it bears less directly on our comfort. We can easily detect the rise in pressure that we experience as we drive down a long steep hill, but most of us would find it difficult, on waking in the morning, to say whether the pressure was higher or lower than it had been when we fell asleep. The humidity can be expressed as relative humidity, wet-bulb temperature, dew point, or water-vapor concentration; any one of these, in combination with temperature and pressure, determines the others.

If we lived on a planet whose atmosphere consisted of a pure gas of uniform composition, we would have only temperature, wind, and pressure to worry about. By the wind I mean the three-dimensional wind, with the strength of an updraft or downdraft appearing as one velocity component. Our own atmosphere, not to mention some other atmospheres in our solar system, is more complicated, in that one of its most important constituents, in our case water vapor, occurs in highly variable concentrations, ordinarily comprising more than two percent of the mass of the atmosphere in the humid tropics, but less than one-tenth of one percent in the colder air at high latitudes or high elevations. Water also occurs as suspended or falling liquid drops and solid particles, so that in reality the atmosphere is not wholly a gas. The variables of the system must therefore include the concentrations of water vapor, which we perceive as humidity, and liquid and solid water, which we observe as denseness of clouds and intensity of rain or snow. It should probably also include the concentrations of such pollutants as dust and smoke.

We could make a case for adding still more quantities, but what makes the atmosphere so complicated as a dynamical system is not so much the proliferation of physical variables as the fact that their values vary from one point to another and not merely from one time to another. To know a single state of the global weather system, we must therefore know the value of each variable at every point. Since there are plainly an infinite number of points in the atmosphere, the system would seem to have an infinite number of variables.

Actually the situation is not quite so bad. On a fine enough spatial scale the weather elements vary rather smoothly, and if two states are nearly alike at each of a sufficiently dense network of well-spaced points, they will be nearly alike at the intervening locations. It is therefore legitimate to treat the atmosphere as a system with a finite number of variables, and to conclude that it is compact. What is not so legitimate is to treat it as a system with a *small* finite number of variables; the number is truly enormous.

How ought we to approach a system of such complexity? Let me present two answers that we might have received at the midpoint of the twentieth century, before the computer had come along and changed everything. Each point of view had its ardent supporters.

Consider first the methods of a subdiscipline that has been known for a century or so as dynamic meteorology, although it might more accurately be called meteorological dynamics. To dynamic meteorologists, the state of the atmosphere consists of the spatial distributions of the temperature, wind, and other weather elements. The dynamicist starts out with the physical laws that govern the behavior of the atmosphere, and usually expresses these laws as mathematical equations. Among them is one of Isaac Newton's laws of motion, familiar to many as "Force equals mass times acceleration," but rearranged for meteorological use as "Acceleration equals force divided by mass," or "Rate of change of velocity equals force per mass." With a knowledge of the state of the atmosphere one can in principle evaluate the force at any point, and thus learn how the velocity of the air passing that point will change as time progresses. The laws of thermodynamics will tell us how the temperature will behave, and other laws will allow us to handle the remaining variables. In short, there is a dynamical basis for forecasting the weather as it evolves, and more generally for treating the atmosphere as a dynamical system.

The synoptic meteorologist would tell a far different tale, regarding the dynamicist's description as grossly incomplete and perhaps irrelevant. Synoptic meteorology is the study of the characteristic structures into which states of the atmosphere can be analyzed. These include meandering jet streams that may encircle the globe in middle latitudes; vortices of subcontinental size, also known as high- and low-pressure systems or simply highs and lows, that travel across the oceans and continents in middle and higher latitudes and bring many of our day-to-day weather changes; smaller and more intense vortices at lower latitudes, known as hurricanes, typhoons, or cyclones according to the ocean over which they originate; towering cumulonimbus clouds with their accompanying thunderstorms and occasional tornados; and small innocuous clouds scattered through an otherwise clear sky. Typical horizontal extents of the structures mentioned are respectively 10,000, 1,000, 100, 10, and 1 kilometers, or a bit more; the list is only a sample. Structures of each type can be counted on to appear again and again, but each individual structure will have its own peculiarities, just as the human race continues, but different people inevitably have their distinguishing personalities. The synoptician's principal tool for identifying and studying the larger structures is the weather map.

The practicing forecaster in precomputer days was effectively an applied synoptic meteorologist. Individual forecasters would learn through their own experience and that of their predecessors how each structure typically develops and moves. They would find that certain peculiarities in a structure signal certain unusual happenings, and they would recognize the telltale signs for the appearance of new structures and the demise of older ones. They would discover, for example, that a high and a low, seemingly heading for the same spot, will retain their identities and simply deflect each other, rather than annihilating each other. In preparing a forecast, a forecaster would most likely construct a prognostic chart, which would be a personal estimate of what the next day's weather map would look like, and he or she would use the chart to infer the coming local weather conditions.

Effectively the forecasting rules are to the synoptician what the physical laws are to the dynamicist. If they could be formulated in such a way as to give a unique prediction in any conceivable situation, they would define an alternative mathematical model, which would constitute another dynamical system.

Why, aside from tradition, didn't some midcentury practicing forecasters opt for the methods of dynamic meteorology? The most likely reason is a practical one; no acceptable weather forecast based primarily on the dynamic equations had ever been produced.

What, then, did dynamic meteorologists have to show for their many years of efforts? As scientists rather than technicians, their interest was directed toward a true understanding of the atmosphere in terms of the physical laws that govern it. They would have been happy to discover why a particular weather pattern with its inevitable peculiarities would evolve as it did, but they were far more interested in why weather patterns vary at all from day to day, or why they are inevitably filled with the large-scale vortices that synoptic meteorologists take for granted.

They would seek to learn what processes would allow these vortices to develop and persist for a while, in the face of the ubiquitous dissipative processes that by themselves would always act to destroy them. By midcentury, they had found many of the answers.

I am not trying to imply that synoptic meteorologists are technicians rather than scientists, nor, for that matter, that technicians are somehow inferior to scientists. It is seldom that a single approach to a problem proves to be the only fruitful one. Synopticians have a keen scientific interest in documenting the properties of the structures that they observe, and in establishing regular relationships between neighboring structures. They are not averse to examining their findings for consistency with the physical laws, but their final conclusions are based more on careful analyses of extensive sequences of weather patterns.

Neither do I wish to suggest that someone who knows all about both dynamic and synoptic meteorology knows all about meteorology. There are numerous other subdisciplines, each with its own group of experts. To name just a couple, one is *cloud physics,* where a fundamental concept is the distribution of the sizes of the drops in a cloud, and where one studies the processes by which tiny suspended droplets and ice crystals become converted into larger drops and particles, which will then fall out as rain or snow. Another is *instrumentation,* where one investigates the strong and weak points of the various instruments via which we have discovered much of what we believe we know about the weather, and where one also designs new instruments in the hopes of gathering hitherto inaccessible information.

Dynamic and synoptic meteorology are not wholly divorced. There have always been some meteorologists who have been outstanding in both subdisciplines. In strong academic meteorology departments, the programs in dynamic and synoptic meteorology tend to be well coordinated. There appear to be other institutions, however, where excellent work in dynamic meteorology may take place in an applied mathematics department, while correspondingly good work in synoptic meteorology may be found in a geography department, but where any communication between the departments is hard to detect. At midcentury, the history of meteorology was marked by both cooperation and contention between the two methodologies.

The Unperformable Experiment

The most direct way to look for chaos in a concrete system, whether it is a simple object sliding down a slope or an atmosphere with its multitude of interdependent structures, is to work with the system itself. If we have released a board and watched or perhaps photographed it on its downward trip, we can easily retrieve it and release it from nearly the same point, to see whether it will follow nearly the same path. Unfortunately for scientific experimentation, but perhaps fortunately for humanity as a whole, we cannot stop the advance of the weather and then reestablish a pattern that has previously been observed, in order to disturb it slightly and then see how rapidly the resulting weather will diverge from the weather that occurred earlier. We can readily disturb the existing weather, perhaps violently by setting off an explosion or starting a fire, or more gently by dropping crystals of dry ice into a cloud—or perhaps even by releasing a butterfly—and we can observe what will happen, but then we shall never know what would have happened if we had left things alone.

What about comparing what happens after we disturb the weather with a forecast of what would have happened if we had not interfered? The forecast is based upon extrapolation from incompletely observed conditions; at best, it can tell us what would have happened if someone had introduced a disturbance similar in magnitude and structure to the observational error—the error in estimating the initial state. If the disturbance that *we* introduce is to tell us anything in addition, it must be large enough not to be swamped by the observational error. However, a disturbance of this magnitude seems hard to produce, when we note that entire thunderstorms may go undetected between observing sites.

Lacking the ability to change the weather to suit our needs, we can wait for what meteorologists call an analogue—a weather pattern that closely resembles one that has previously been observed—in order to see how closely the behavior following the second occurrence resembles that following the first. This method also fails; even though the atmosphere seems to be a compact system—one in which pairs of analogues must eventually occur—good analogues on a global scale have not been found within the few decades that global weather conditions have been recorded. Patterns that are much alike over regions of continental size are sometimes observed, but, when these fail to develop similarly, they may do so because dissimilar weather structures have moved in from distant regions, rather than because of any sensitivity to small local differences.

There remains the reasonably well established observation that weather variations are not periodic. Of course they have periodic components, the most obvious ones being the warming and cooling that occur with the passage of the seasons of the year or the hours of the day. Careful measurements have also detected weak signals with a lunar period, probably gravitational effects, and there is virtually no limit to the number of periods that investigators have *claimed* to have discovered. Some of these have been stated to several decimal places. Nevertheless, if we take an extended record of temperature or some other weather variable and subtract out all verified or suspected periodic components, we are left with a strong irregular signal. Migratory storms that cross the oceans and continents are still present in full force. These are presumably manifestations of chaos.

Since good analogues of global extent have yet to be discovered, we cannot with certainty rule out the possibility that, when one finally does appear, the subsequent weather will repeat the earlier sequence. That is, the atmosphere may really be behaving periodically, with a period whose length exceeds that of any weather records. We are left with the strong impression that the atmosphere is chaotic, but we would like additional evidence.

Voices from Dishpans

It is fairly easy to construct a scale model of a bumpy slope and observe the descent of a ball or some other object. Modeling a planetary atmosphere in the laboratory is another matter. We might think of letting a fluid fill the space between two concentric spheres. The inner sphere could represent the planet, while the outer one could take the place of gravity to the extent of preventing the fluid from leaving the planet, but how could we then introduce a force that would simulate gravity within the fluid by always being directed toward the planet's surface?

The pioneers in laboratory modeling were already well versed in dynamics. Dynamic meteorologists have long been accustomed to simplifying their equations before putting them to use. Sometimes the simplifications are merely deletions of terms that appear to be inconsequential, but equally often they consist of omitting or significantly altering certain physical features or processes. Thus, effectively, they replace the atmosphere by a different atmosphere, which Napier Shaw described in the early twentieth century, in his four-volume treatise *Manual of Meteorology,* as a fairy tale, but which today we would call a model. The dynamicist assumes, or at least hopes, that the weather in a make-believe atmosphere will be more or less like the real weather in its gross features, and will differ mainly in minor details.

The equations expressing the laws that govern evaporation and condensation of water in the atmosphere are rather awkward, while those governing the transformation of a cloud composed of tiny suspended water droplets into a rain cloud are even more forbidding, and dynamicists often work with model atmospheres that are devoid of water in any form. Likewise, many of them have an undisguised aversion to spherical geometry, and their atmospheres may flow over a flat rotating earth instead of one whose surface is curved. Any dynamic meteorologist who could explain the development and persistence of large-scale vortices in the fairyland of dry atmospheres and flat earths would feel that he had completed the major part of the work of solving the real-world problem.

Laboratory modelers found it quite acceptable to build into their apparatus the same distortions that dynamic meteorologists traditionally built into their equations. The first experiments to bear fruit were designed by Dave Fultz at the University of Chicago. After a number of trials, he settled upon a cylindrical vessel partly filled with water, placed on a rotating turntable, and subjected to heating near the periphery and cooling near the center. Figure 28 is a schematic oblique view of his apparatus. The bottom of the container is intended to simulate one hemisphere of the earth's surface, the water is intended to simulate the air above this hemisphere, the rotation of the turntable simulates the earth's rotation, the heating and cooling simulate the excess external heating of the atmosphere in low latitudes and the excess cooling in high latitudes, and gravity simulates itself. Fultz had hoped that the motions that developed in the water might resemble the large-scale wind patterns in the atmosphere.

Originally the edge of the container extended beyond the rim of the turntable, and the heating was accomplished by a fixed Bunsen burner, while exposure to room temperature was supposed to take care of the

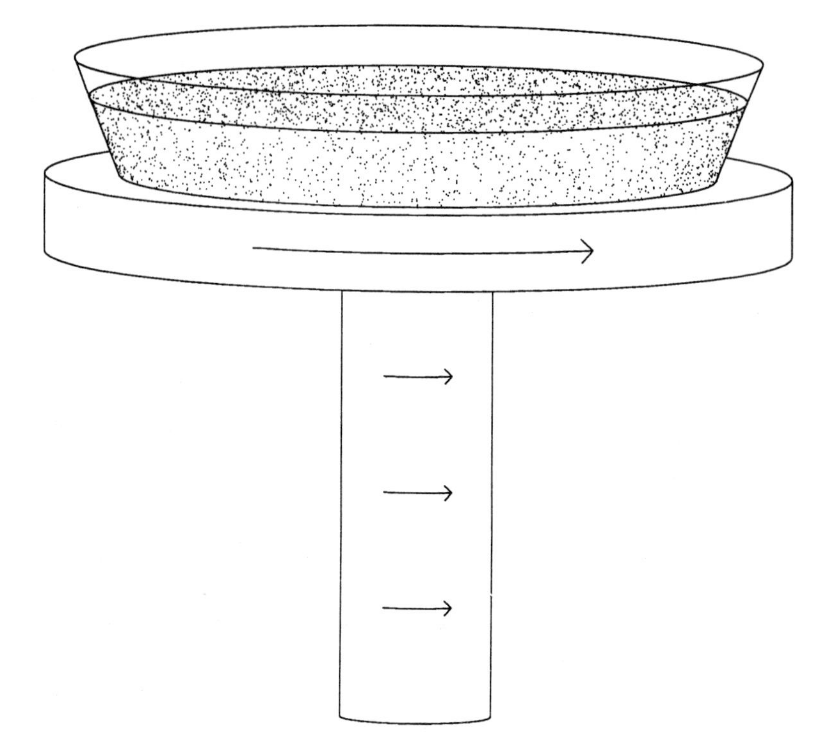

Figure 28. A schematic oblique view of the apparatus used by Fultz in the dish-pan experiments. The arrows indicate the direction of rotation of the turntable. Heating is applied at the rim of the container and cooling is applied at the center.

cooling. This setup was soon supplanted by more easily controlled heating coils arranged around the periphery of the container, while the cooling was sometimes accomplished by an upward jet of cold water through a hole in the turntable. Flow at the upper surface, which was intended to simulate atmospheric motion at high elevations, was made visible by a sprinkling of aluminum powder. A special camera that effectively rotated with the turntable took time exposures, so that a moving aluminum particle would appear as a streak, and sometimes each exposure ended with a flash, which would add an arrowhead to the forward end of each streak. Flow deeper within the fluid could be detected by injecting a dye, and thermometers were often inserted to record the temperature fluctuations that were expected to accompany the passage of the simulated weather structures. The turntable generally rotated counterclockwise, as does the earth when viewed from above the north pole.

The collection of components cost about forty thousand dollars—a fair sum for 1950—but the central component—the container—was an ordinary dishpan purchased for a dollar or two, and the work became known as the dishpan experiments.

Fultz had assumed that it would make little difference whether the working fluid was a gas or a liquid, and water was certainly the simplest choice. The water possessed no impurities to simulate real atmospheric water vapor and clouds, and the bottom of the dishpan was essentially flat, with nothing to distinguish between oceans and continents. Dynamicists who might have been criticized for omitting the water in the atmosphere and the curvature of the earth could have claimed that they were really trying to model the dishpan experiments.

Because everything in the experiments was arranged with perfect symmetry about the axis of rotation, at least within the limits of experimental control, one might have anticipated that the resulting flow patterns would also be symmetric, looking perhaps like the one shown schematically in Figure 29. Fultz was hoping for something more like Figure 30, with a meandering jet stream and an assemblage of vortices typical of the atmosphere.

He got more than he bargained for. Both flow patterns appeared, the choice depending upon the speed of the turntable's rotation and the intensity of the heating. In brief, with fixed heating, a transition from circular symmetry would take place as the rotation increased past a critical rate. With fixed, sufficiently rapid rotation, a similar transition would occur when the heating reached a critical strength, while another transition back to symmetry would occur when the heating reached a still higher critical strength. The experiments proved to be repeatable, producing transitions at the same combinations of values of rotation and heating when run again.

In the early experiments, the flow that was asymmetric appeared to be irregular also, changing continuously from one pattern to another, much as the real atmosphere changes. We now recognize Fultz's transitions as classical bifurcations, between a steady system, whose attractor consists of a single point in phase space, and an unsteady, apparently chaotic one, whose attractor should be composed of an infinite complex of multidimensional manifolds.

In England, meanwhile, Raymond Hide was experimenting at Cambridge University with a somewhat similar apparatus. It differed mainly

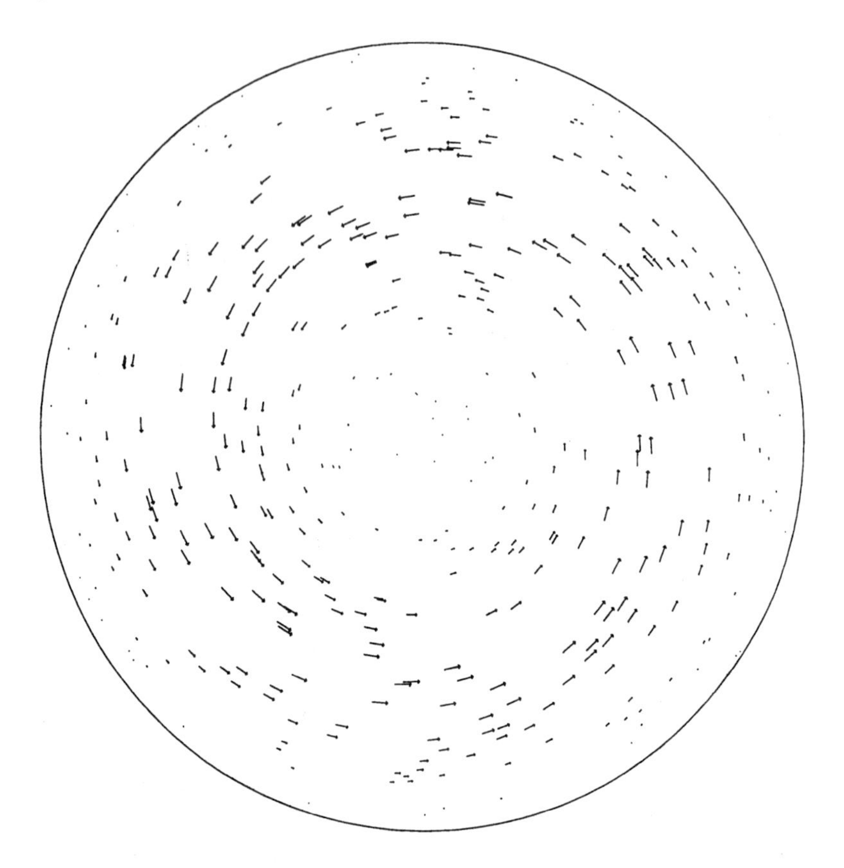

Figure 29. A schematic view of symmetric flow at the upper surface of the water in the dishpan.

in that the fluid occupied a ring-shaped region between two concentric cylinders instead of the interior of a single cylinder. Hide found similar transitions between symmetric and asymmetric flow, but, possibly because of the restrictive effect of the inner cylinder, the asymmetric flow was often regular, and would consist of a chain of apparently identical waves, which would travel around the cylinder without changing their shape. Here was a dynamical system with a one-dimensional attractor—a circle in a suitably chosen phase space—the separate states being distinguished only by the longitudes of the waves.

Hide also discovered a remarkable regular phenomenon, which he called *vacillation*. Here also a chain of identical waves would appear, but,

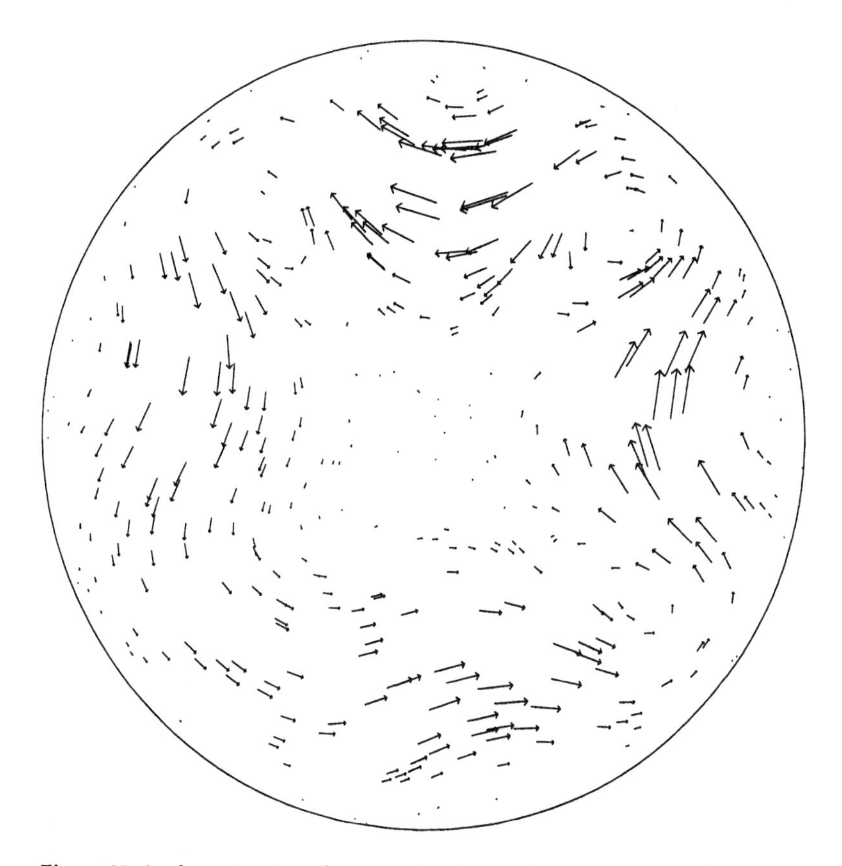

Figure 30. A schematic view of asymmetric flow at the upper surface of the water in the dishpan.

as they traveled along, they would alter their shape in unison in a regular periodic fashion, and, after many "days"—many rotations of the turntable—they would regain their original shape and then repeat the cycle. Here the system had a two-dimensional attractor, the two varying quantities being the longitudinal phase of the waves and the phase of the vacillation cycle.

Hide was not a meteorologist, although he has since become one of the leading dynamicists in the meteorological community, and he was actually attempting to model the motions in the earth's magnetic core, but as an all-arounder he quickly saw the relevance of his experiments for the atmosphere, and noted the resemblance between his vacillation cycles and the frequently seen fluctuations between intervals of strong and weak westerly winds. He and Fultz soon traded information. In due time Fultz produced both uniform wave motion and vacillation in the dishpan, and Hide found that his own apparatus would support irregular behavior.

Other scientists soon took up the laboratory modeling, although the number remained small compared, for example, with those who continued to favor equations. By using such components as a discarded phonograph turntable, Alan Faller managed to reproduce the essence of Fultz's early experiments at a cost of four dollars. Subsequently, at the Woods Hole Oceanographic Institution, he built a "dishpan" eight feet in diameter, and was able to produce cold fronts and warm fronts—narrow zones separating extensive air masses, or, in the experiments, water masses, with contrasting temperatures. Fronts are ubiquitous features of sea-level weather maps.

Figure 31 shows a vacillation cycle as captured photographically by Fultz. The separate pictures were taken at intervals of four "dishpan days." In the first picture, a meandering circumpolar jet stream, identifiable by the longest bright streaks, appears to be made up of five virtually identical waves. The waves proceed to change their shape, and after eight days they have become transformed into nearly circular vortices. The vortices subsequently elongate, and by sixteen days the pattern, not shown, has become virtually indistinguishable from the initial one.

Figure 32 shows two photographs of the dishpan, one "day" apart, during an irregular and presumably chaotic regime. Accompanying them are Fultz's streamline analyses, based on the photographs. A nearly circular vortex below the center may be seen to elongate considerably as it propagates "eastward." The vortex rotates counterclockwise, and, like its counterparts in the northern hemisphere of the real atmosphere, it is a true low-pressure center.

In Figure 33 the first streamline analysis has been inverted so as to simulate southern-hemisphere flow, and it is compared with an actual upper-level southern-hemisphere weather map, containing approximate streamlines. The patterns are not superposable, but qualitatively they are so much alike that they might almost have been selected from the same attractor.

The implications of the laboratory experiments are profound. Structures such as jet streams, traveling vortices, and fronts appear to be basic

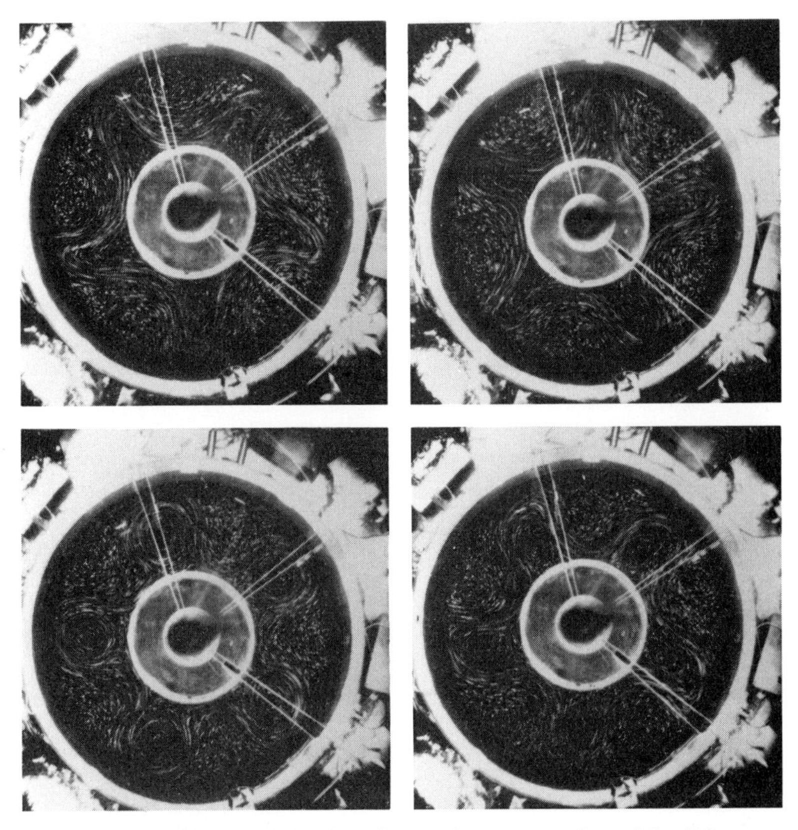

Figure 31. Streak photographs of the flow at the upper surface of the dishpan, at four phases of a vacillation cycle. The upper right, lower left, and lower right patterns follow the upper left by four, eight, and twelve rotations. The pattern after four more rotations, not shown, is almost indistinguishable from the first. Photographs by Dave Fultz.

features of rotating heated fluids, and are not peculiar to atmospheres. Efforts in dynamic meteorology had not always been success stories, and it had been proposed at times that the failures might result from using the wrong dynamics—possibly from being unaware of some strange force. The similarities and differences between the atmosphere and the experiments strongly suggest that the principal causes of the gross behaviors are to be found in forces and processes that influence both systems—gravity, rotation, and differential heating—while such properties as compressibility, which air possesses but water does not, are second-

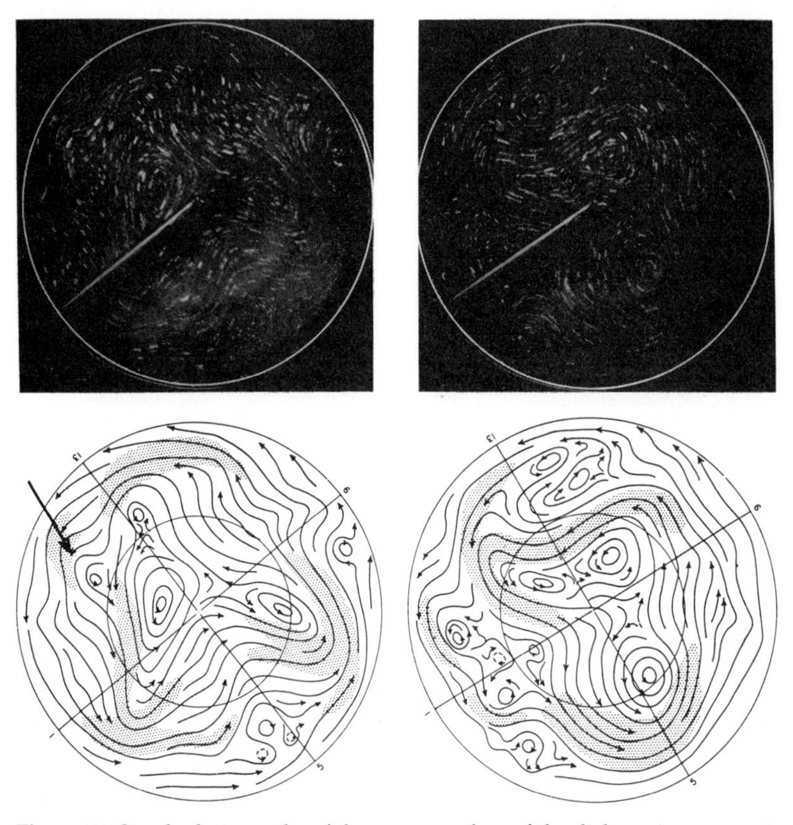

Figure 32. Streak photographs of the upper surface of the dishpan in an experiment revealing chaotic behavior, accompanied by streamline analyses based on the photographs. The right-hand patterns follow those on the left by one rotation. Photographs and analyses by Dave Fultz, reproduced by permission of the American Meteorological Society.

ary. If a mysterious force is at work in the atmosphere, it would have to be at work in the dishpan also.

Finally, it may be a prehistorical accident that our day is about twenty-four hours long, instead of several times longer. If so, it may be an accident that our atmosphere behaves like a rapidly rotating dishpan instead of a slow one, fluctuating chaotically instead of regularly, and that our weather is rather unpredictable instead of continually executing a monotonous cycle, or perhaps not changing at all except for the slow advance of the seasons.

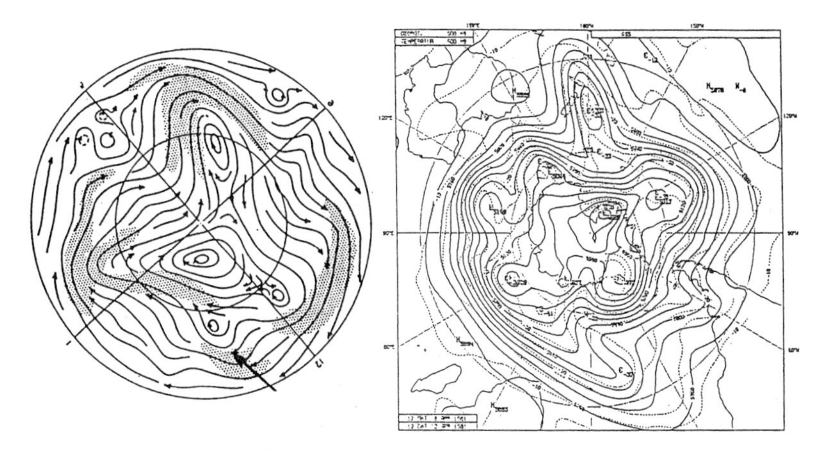

Figure 33. The left-hand streamline analysis from Figure 32, inverted so as to simulate southern hemisphere flow, compared with a real-world height-contour analysis of the 500-millibar surface over a portion of the southern hemisphere. At this surface, whose average height is close to 5.5 kilometers, the height contours closely resemble streamlines.

The Five-Million-Variable Dynamical System

By far the strongest evidence for a chaotic atmosphere has come from mathematical models. Strictly speaking, these models are what dynamicists have been using since the dawn of dynamic meteorology, but more recently a "model" has generally meant a system of equations arranged for numerical solution on a computer. The history of the more realistic models of this sort is essentially the history of the use of the dynamic equations for weather forecasting.

The story opens in Norway in the early twentieth century, when Vilhelm Bjerknes, considered by some to have been the all-time great meteorologist, proposed that the weather-forecasting problem was simply the problem of solving the system of equations representing the basic laws, using a set of simultaneous weather observations as the initial state. He maintained that the equations were known, but recognized that there was no practical method of solving them. It was Bjerknes who, many years later, championed the idea that the reason that vortices and other structures of continental or subcontinental size must be present in the atmosphere is not the dynamic impossibility of a flow pattern without them, which would look like symmetric flow in the dishpan, and would constitute a state of equilibrium. Rather, it is the instability of such a pattern with respect to inevitable disturbances of large horizontal extent but small amplitude. These disturbances would proceed to intensify and then persist as part of the complete pattern.

The next chapter begins in England during World War I, when the versatile scientist Lewis Richardson, who was undaunted by the formidable nature of the equations, attempted to solve them numerically. In his procedure, he replaced the continuous distributions of pressure, wind, and other quantities, which in any event could be estimated only by interpolating between reports at weather stations, by the values of these quantities at a regular grid of points. The gradients of these quantities—the rates at which they varied horizontally—were then approximated by differences between values at adjacent grid points.

As a Quaker, Richardson objected to armed combat, but he had no fear of the front, and was happy to serve during the war as an ambulance driver. He brought with him an extensive set of weather data for one selected day, and between shifts he would perform the thousands of additions, subtractions, and multiplications needed to produce a single six-hour forecast for an area no larger than Europe. His predicted weather pattern not only turned out to be wrong, but did not even resemble any pattern that had been seen before. Richardson correctly attributed his failure to inadequacies in the initial wind measurements, although subsequent analysis has shown that his procedure would have produced serious although less drastic errors even with perfect initial data.

Imagine an enormous creature from outer space that swoops down close to the earth, reaches out with a giant paddle, and stirs the atmosphere for a short while before disappearing. Wholly aside from the possibly disastrous effect upon the living beings of the earth, what will be the likely effect on the weather?

Air that has simply been moving around a low pressure system, for example, as it normally does, may be left moving predominantly into it. The low will rapidly fill, soon becoming a high, after which the now piled-up air will surge outward, leaving a deep low, into which air will rush a second time before rushing out again. The precise chain of events will be further complicated by the ever-present deflecting effect of the earth's rotation. Rather similar events will take place at locations where the creature has left the air moving out of a low, or into or out of a high. In short, there will be violent fluctuations of pressure and accompanying fluctuations in the wind. Theory indicates that the period of an oscillation—a change from low to high to low again—will be comparable to one day.

The atmosphere has its own method of getting rid of any such fluctuations; otherwise they might be a part of our normal weather. Mechanical and thermal damping play an essential role in their removal. After a few weeks, the oscillations will be hardly detectable and the weather will be back to normal, although the particular sequence of weather patterns will undoubtedly not be the one that would have developed without the disturbance. Stated otherwise, the normal weather patterns that occur day after day belong to the attractor of the global weather system. The alien creature will produce a new state—think of it as an initial state—that is well removed from the attractor, but, as in any dissipative dynamical system, the transient effects will ultimately damp out, and normal behavior will resume.

Now imagine that Richardson had wished to use his numerical procedure to discover what would happen if such a creature should pay a visit. He could have done no better than to do what he actually did. Observations of the weather such as those that he used, and interpolations to standard geographical locations, are fraught with small errors. The true state of the atmosphere, and the state as Richardson could best estimate it, differed in much the same way as the states of an atmosphere before and after being stirred. The true state belonged to the attractor, and the estimated state did not. Inevitably Richardson predicted the violent oscillations that his assumed initial state demanded.

In his monumental book *Weather Prediction by Numerical Process,* completed in 1922, Richardson presented his procedure in full detail, and discussed his forecast. He ended by envisioning a weather center where sixty-four thousand people working in shifts could produce a weather forecast more rapidly than the weather itself could advance. The one feature that he failed to envision was the device that within half a century would be working as rapidly as sixty-four thousand people.

Following Richardson's efforts, the general attitude toward numerical weather prediction became pessimistic. Many prominent meteorologists seriously doubted that wind observations could ever become accurate enough to suppress the spurious oscillations. Those who felt otherwise tended to be discouraged by the sheer magnitude of the needed computations.

Fortunately one of the optimists was the Swedish scientist Carl-Gustaf Rossby, a dynamic meteorologist in the literal as well as the technical sense, and certainly another candidate for the title of all-time great meteorologist. One of his contributions was the suggestion that the key to understanding the atmosphere was to be found in the wind instead of the pressure. A low-pressure system is also a spinning vortex, and, although the barometer provides the easiest and most accurate means of detecting and mapping the structure, the wind pattern may exert the greater influence on its behavior, with the pressure serving largely as an indicator.

As the middle of the century approached, the renowned mathematician John von Neumann became interested in designing automatic computers and applying them to involved problems. Although not a meteorologist, he recognized the weather-forecasting problem as ideal for his needs. Soon he went about assembling a group of meteorologists and other scientists to work on the numerical forecasting problem, at the Institute for Advanced Study in Princeton, New Jersey. In addition to the largely computational matters to be faced, there remained the problem of the spurious fluctuations.

The states that the atmosphere assumes as the weather progresses are all supposed to be restricted to the attractor of the global weather system; any transient effects should have disappeared long ago. If some system of equations is to be used to produce short-range forecasts, say one or two days in advance, it must handle the states that are on the attractor, or the best approximations to these states that it is able to depict, in approximately the way that the atmosphere handles them. On the other hand, there is no need for it to be able to deal properly with states that are not on the attractor, since these will never arise.

One member of von Neumann's group who recognized this situation was the then-young meteorologist Jule Charney, later to be recognized as still another possible all-time great. Before arriving in Princeton, Charney had become acquainted with Rossby. Starting from Rossby's ideas as to the importance of the wind, he had managed to construct a system of equations that effectively failed to distinguish between unrealistic weather patterns in which strong oscillations would have been expected to develop, and slightly different but more realistic ones in which they would not, and, with either type of pattern as an initial state, would predict that the oscillations would not arise. His system could not have detected a visit from the creature from outer space. Effectively the new equations filtered out the oscillations, and later were sometimes called the filtered equations, while the more nearly exact equations that Richardson had used became known as the primitive equations. With various modifications that rendered them more adaptable to computation, the filtered equations became the basis for the first experimental series of numerical weather predictions. The story of these early efforts has been aptly recounted by both Philip Thompson and George Platzman, two of the original participants.

By the middle fifties, the moderate success of the forecasts, even though they did not match up to the ones turned out by experienced synoptic forecasters, led to the introduction of numerical forecasting as a part of the operational procedure of various national weather services. At the very least, synoptic forecasters now had, in addition to everything that they had formerly used, the information that "this is what the computer says will happen." They could use or reject this information as they saw fit. As the years advanced, forecasters came to rely more and more on the numerical product.

As dynamical systems, the new models were rather peculiar. They did not possess attractors that resembled the attractor of the real atmospheric system. If they had been used to make long-range forecasts, say a month in advance, they would have produced weather patterns quite unlike anything seen in nature. Indeed, in the earliest forecasts the external forcing and internal dissipation were completely omitted from the equations, on the grounds that, no matter how important they might have been in bringing about an initial state, their added influence during the next day or two would be minor. Thus, aside from any changes that the numerical scheme might have introduced, the models conserved total energy, and, like other Hamiltonian systems, possessed no attractors at all.

An outgrowth of numerical weather prediction that recognized this shortcoming was global circulation modeling. The equations used were much like those already used in short-range prediction, but, as the name suggests, the area to which they were applied covered the whole globe, or at least most of one hemisphere, rather than a more restricted region. The purpose of the new models was to produce simulated weather whose long-term behavior was realistic in as many respects as possible, rather than to make forecasts, and the initial conditions were often purposely chosen not to look like real weather patterns, in the hopes that reasonable patterns would soon develop. Stated otherwise, it was hoped that the new models would possess realistic attractors. Needless to say, external forcing and internal dissipation had to be included.

The prototype global circulation model was constructed by Norman Phillips, who had been working closely with Charney at the Institute for Advanced Study. His dynamical system had 450 variables. He extended his solution for one month, and produced a meandering jet stream and traveling vortices before he encountered computational problems. Subsequently he succeeded in diagnosing the computational difficulty and prescribing a cure, thus paving the way for the countless models that were to follow.

During the sixties it became apparent that the filtered equations, which had made numerical forecasting possible with the early computers, were not going to produce forecasts of the quality that some had hoped for. With the advent of more powerful computers, some meteorologists turned their attention back to the primitive equations. The solution to the problem of the violent oscillations, which had led to the rejection of the primitive equations a decade earlier, turned out to be surprisingly simple in concept, although not so easy to implement.

The initial patterns of wind and perhaps pressure, as interpolated from observations, are inaccurate in any case; otherwise they would not give rise to oscillations so much stronger than those observed in nature. Why then shouldn't we tamper with these patterns a bit, at the risk of making them slightly more inaccurate, to produce *new initial states* from which oscillations *cannot* develop, as an alternative to using *equations* that will not *predict* that oscillations will develop? After extensive research, several methods of making the needed adjustments were devised; improvements are still being introduced. The tampering or adjusting process, known as *initialization,* is now a standard part of every routine forecasting procedure that uses the primitive equations. Let us note that initialization need not produce the correct initial state; it simply produces one that might be correct instead of one that cannot be. Ideally, it will move the observed state to some nearby state on the attractor.

By the seventies, global circulation modelers were also turning to the primitive equations. As the years advanced and computers became ever more powerful, the distinction between global circulation models and numerical forecasting models tended to disappear. Operational forecasting centers could now afford to use models that covered the globe, or at least a hemisphere, and, with increasing interest in predicting several days ahead, during which time storms could move in from distant areas, there was good reason to do so.

The big model with which I am personally most familiar is the operational model of the European Centre for Medium Range Weather Forecasts in Reading, England. The Centre is a joint venture of the weather services of more than a dozen European nations. As its name implies, its mission is to produce forecasts at medium range, which in this case has meant up to ten days ahead. The scientists who have worked there, either directly with the model or on problems relevant to its construction and performance, have included not only representatives from the member nations but visitors from around the world.

The model itself is global, and, like most large models today, is based on the primitive equations. I should probably not call it *the* model, because, during the ten years or so that I have intermittently worked with it, it has been frequently subjected to modifications aimed at improving its performance. As of 1985, it possessed three physical quantities—temperature and two wind components—defined at each of nineteen elevations, and a fourth quantity—water-vapor content—defined at all but the high elevations. Pressure was explicitly defined only at the lowest level, since pressures at higher levels could be inferred from those at lower levels when temperatures and humidities at intervening levels were known. Other auxiliary variables such as soil moisture were defined where appropriate. Each physical quantity at each level was for practical purposes specified at a grid of more than 11, 000 points, spanning the globe. This produced a total of some 800, 000 variables.

As if these were not enough, as of late 1991 the resolution was doubled in both the latitudinal and longitudinal directions, producing effectively 45, 000 grid points, while the number of standard elevations was increased to 31. This produced a model with five million variables. Of such stuff are modern global circulation models made.

Lest a system of 5, 000, 000 simultaneous equations in as many variables appear extravagant, let us note that, with a horizontal grid of less than 50, 000 points, each point must account for more than 10,000 square kilometers. Such an area is large enough to hide a thunderstorm in its interior. I have heard speculations at the Centre that another enlargement of the model is unlikely to occur soon, and this evidently means that not all significant weather structures will soon be resolved.

What about chaos? Almost all global models, aside from the very earliest, have been used for predictability experiments, in which two or more solutions originating from slightly different initial states have been examined for the presence of sensitive dependence. Interest has not been so much in chaos itself as in the feasibility of producing extended-range forecasts, particularly at the two-week range.

Almost without exception, the models have indicated that small initial differences will amplify until they are no longer small. There is even good quantitative agreement as to the rate of amplification. Unless we wish to maintain that the state-of-the-art model at the European Centre, and competitive models at the National Meteorological Center in Washington and other centers, do not really behave like the atmosphere, in spite of the rather good forecasts that they produce at short range, we are more or less forced to conclude that the atmosphere itself is chaotic.

The Consequences

The possibility that an object may slide chaotically down a slope is largely a matter of academic interest. Chaos in the atmosphere has farther-reaching consequences.

The most obvious effect is the limitation that it imposes upon our ability to forecast. Imagine that you are a computer-age weather forecaster, and that instead of making just one extended-range forecast you have decided to make a dozen or so. You take a dozen estimates of the initial state that are more or less alike, differing from each other by no more than the typical uncertainty in estimating the true initial state: temperatures at some locations might differ by a degree or so, while wind speeds might vary by two or three knots. To fend off anticipated spurious oscillations, you apply the initialization procedure to each state. When you make a two-week forecast from each state, using the best approximation to the true physical laws that your computer can handle, you will find that, because of sensitive dependence, the dozen predictions are not much alike. If you have no basis for saying which, if any, of the dozen initial states is correct, you will have no basis for saying which of the predictions should become the official forecast.

The process that you will have carried out is not somebody's wild fantasy. It shows signs of becoming the forecasting procedure of the future, when computers have become still more powerful. It is known as *Monte Carlo* forecasting. It takes its name from the famous gambling resort because the original idea was that, out of a virtually infinite collection of eligible initial states, a few should be chosen at random, although it now appears that, if the procedure becomes operational, the states may be chosen systematically. Monte Carlo methods have in fact been used in numerous fields almost since the advent of computers as a means of generating statistical distributions.

The Monte Carlo procedure can give some idea of the degree of confidence to be put in a particular day's forecast. If the separate forecasts show little resemblance to each other, the confidence will be low, whether one of the forecasts is selected arbitrarily as the official one, or whether some average is used. If the forecasts are much alike, any one of them is likely to be fairly good.

What is the basis for choosing two weeks as a time after which the forecasts might differ significantly? That story goes back to the early 1960s, when preparations were under way for the Global Atmospheric Research Program, the international effort to obtain world-wide observations of a higher quality than previously known, and to extend our knowledge of atmospheric dynamics so that the new observations might be put to optimum use. Among the original aims of the program was the production of good two-week forecasts. Such a vast program obviously would require vast funding, and "aims" should perhaps be viewed as a euphemism for "selling points."

It was just at that time that the possibility of sensitive dependence in the atmosphere's behavior was beginning to be recognized by meteorologists. Jule Charney, who was one of the leaders in organizing the program, became concerned that two-week forecasting might be proven impossible even before the first two-week forecast could be produced, and he managed to replace the aim of making these forecasts with the more modest aim of determining whether such forecasts were feasible. In 1964, a special conference held in Boulder, Colorado, was attended by a wide assortment of dynamicists, synopticians, and other meteorologists, including all the then-active global circulation modelers. The agenda included scientific papers presented by representatives of ten nations, and Charney talked about the possibility of chaotic behavior.

Between sessions, however, when the real work of such conferences generally takes place, he managed to persuade all of the global-circulation modelers to use their models to perform numerical experiments in which pairs of forecasts originating from slightly different conditions would be examined for sensitive dependence. From these experiments, performed in the ensuing months, Charney's committee concluded that a reasonable estimate of the average doubling time for small errors in the temperature or wind pattern was five days.

The doubling time soon acquired the status of a standard measure of predictability. If we have a fair idea of the size of typical errors in estimating the initial state, and if we have decided how great an error we can tolerate in the forecast, we will know how many doublings are acceptable, and, if we also know the typical doubling time, we can calculate the range of acceptable predictability. This range should then be adjusted upward, since errors typically grow less rapidly after they have become moderately large. The five-day doubling time seemed to offer considerable promise for one-week forecasts, but very little hope for one-month forecasts, while two-week forecasts seemed to be near the borderline.

What typically happens when a more powerful computer becomes available to the meteorological community is that larger mathematical models are built, so that a one-day or a one-week forecast takes about as long to produce as it did before. The enlargements generally entail increases in horizontal and vertical resolution, but they may also involve more realistic formulations of certain physical processes, such as the absorption and emission of radiation by the atmosphere, or the flow of air over mountainous terrain. One specific enlargement in the sixties was the change from filtered to primitive equations. With the new models came new predictability experiments, and by 1970 the typical doubling time seemed to be closer to three days than five. By the early eighties, the European Centre model and other models had reduced the time to about two days; this estimate still stood in 1990. Thus, although it had become fairly well established that two-week forecasts showed a slight edge over pure guesswork, scenarios in which the locations and intensities of migratory storms were predicted two weeks ahead appeared less and less realistic.

Some promise for further improvement in forecasting has come from the observation that, with the European Centre model, differences between two forecasts that start from different states regularly amplify more slowly than differences between either forecast and the weather that actually transpires. If the model perfectly represented the physical laws, the rates of amplification should all be the same. Improvements must therefore still be possible. Computations indicate that, if the present model is correctly estimating the atmosphere's doubling time, a perfect model should produce three-day forecasts as good as today's one-day forecasts, which generally are good; one-week forecasts as good as today's three-day forecasts, which occasionally are good; and two-week forecasts comparable to today's one-week forecasts, which, although not very good, may contain some useful information. This is the optimistic view; one can always take the alternative view that, since the present models are not perfect, the appropriate doubling time may be even less than the estimated two days.

Let us take another look at the calculated doubling times. First of all, they are properties of the models that have been used to compute them. As a property of the real-world system, a two-day doubling time can at most refer to a doubling time for structures that are resolved by the models—jet streams, temperate latitude vortices, and perhaps tropical storms. Structures that are not regularly resolved either by a global observational network or by the computational grid of a global model have an important influence on the resolved structures; thunderstorms, and to some extent less intense showers, are effective in altering the global temperature and humidity patterns by carrying heat and water from low to high elevations. Failure to include these effects in a model leads to inferior forecasts. The larger-scale pattern tends to be indicative of the presence or absence of significant smaller-scale structures, and the standard procedure is to estimate, at each point, the most probable effect of the smaller scales. This procedure, know as *parameterization,* has been the subject of entire conferences. It is still one of the less realistically formulated aspects of large models, and amendments or alternative schemes are continually being introduced.

If the models could ever include so many variables that individual thunderstorms and other smaller-scale structures would be properly represented, and parameterization would no longer be needed, it would be totally unreasonable to expect that errors in the details of these structures would require two days to double. Individual thunderstorms typically last only a few hours, and, with an assumed two-day doubling time, the error growth during those hours would be nearly imperceptible. Since a thunderstorm can in reality easily double its severity in less than one hour, we should expect that the difference between two rather similar thunderstorms would double just as rapidly.

If this is the case, the outcome would be that, after several hours, forecasts of the details of small-scale structures would be no better than guesswork, and subsequent representations of their effects on the larger scales would be no better than parameterization. In other words, if we could use such a model with its unbelievably high resolution for perhaps the first half day, we might as well return to one of today's models for the remainder of the forecast. The implication is that introducing such impossibly high resolution would increase the range of practical predictability by only a few hours. As a corollary, it appears that coming improvements in forecasting may have to come from better numerical representations of the structures that are supposedly already resolved, or better formulations of some of the physical processes. The apparent drop in returns with continued increases in resolution has led some forecasters to propose that the anticipated additional computer power in the middle nineties can be more advantageously used to carry out some Monte Carlo procedure.

With all these obstacles around, it may surprise us to learn that within our chaotic atmosphere there are certain weather elements at a few locations that can be rather accurately predicted not just two weeks but two months or even two years ahead. The most spectacularly predictable of these are the winds at high levels in equatorial regions, which are dominated by the so-called quasi-biennial oscillation, first recognized by Richard Reed of the University of Washington. Since the cataclysmic eruption of Krakatau west of Java in 1883, it had been common "knowledge" that the winds at 20 or 25 kilometers above the equator blew from the east; a cloud of volcanic dust had even been observed to circle four times about the globe.

In the 1950s, when sporadic equatorial balloon soundings first reached high enough elevations, a few meteorologists noted that the "normal" so-called Krakatoa easterlies were sometimes missing. I was fortunate enough to be present at the meeting in 1960 when Reed announced his findings, and I could see members of the audience shaking their heads as he maintained that at these heights the equatorial winds would blow continually from the east for about a year, and then from the

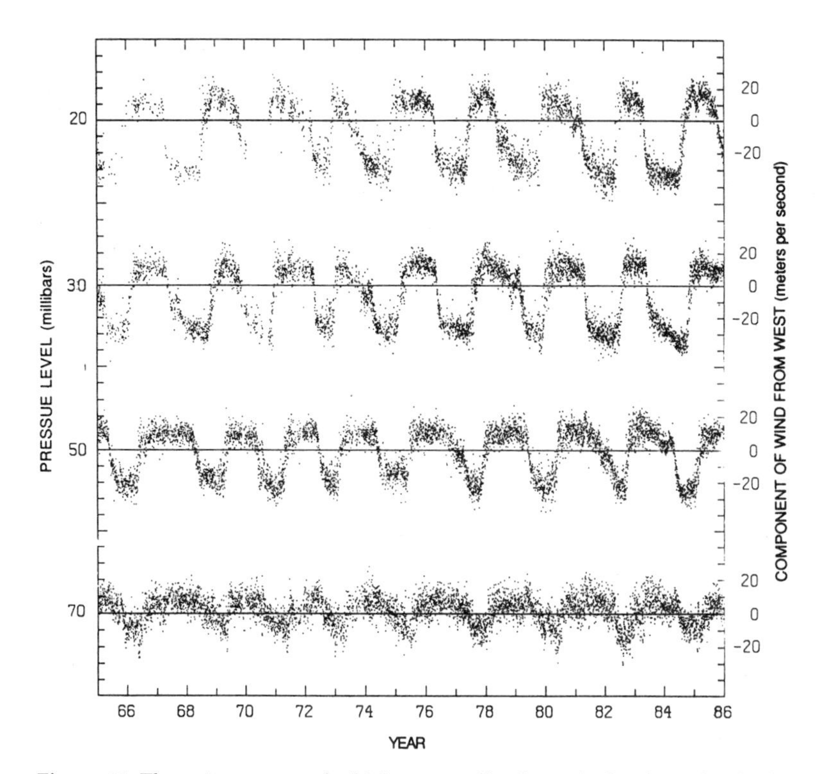

Figure 34. The points, some of which are too closely packed to be individually distinguishable, show daily values of the eastward component of the wind at the 70-, 50-, 30-, and 20-millibar surfaces over Singapore, from 1965 through 1985 as indicated by the scale at the base. The solid lines are zero-lines. Values above zero indicate winds from the west. The approximate two-year periodicity is evident.

west for a year, and then from the east again for another year, and that, if Krakatau had blown up a year earlier or later, the meteorological language would have had the term "Krakatoa westerlies."

The subsequent years have fully confirmed his claims. In Figure 34, the plotted points show daily observed values of the eastward component of the wind above Singapore, one degree north of the equator, at the four standard pressure levels of 70, 50, 30, and 20 millibars—the pressure is about 1000 millibars at sea level. Above Singapore, these pressures are reached at elevations that fluctuate a few hundred meters about averages of 18.6, 20.6, 23.8, and 26.3 kilometers—roughly twice the height at which commercial jets typically fly. The sequences extend from the beginning of 1965, when the sounding balloons released at Singapore first regularly reached the high levels, to the end of 1985, and they cover nine complete cycles. No smoothing or averaging has been performed, so that the points represent the values that one would attempt to forecast when forecasting for particular moments. Although there are always some day-to-day fluctuations, by far the stronger part of the signal at the upper three levels consists of the oscillation itself, and it is apparent that forecasts with reasonably small errors, for the winds on most of the individual days a cycle or two in advance, can be produced by subjectively extrapolating the phase, and predicting conditions that are average for that phase. As with any other forecasts, these ones will sometimes fail completely, particularly for the times when the rather sharp transitions between westerlies and easterlies will be occurring. Note that the phases propagate downward, taking about a year to descend from 20 to 70 millibars.

The period of oscillation is not exactly two years, as had been conjectured when fewer cycles had been observed, and the separate cycles are not of identical length, so that the oscillation is presumably chaotic, and its phase cannot be predicted decades in advance. Yet the chaos is characterized by an entirely different time scale from that of storms of continental size, just as these storms have a different time scale from thunderstorms. Perhaps the principal lesson is that we still have much to learn about what can happen in chaotic dynamical systems with many interconnected parts.

Other conclusions as to the consequences of chaos in the atmosphere are more speculative, and result from comparing the real world with guesses as to what the world would be like if the weather were not chaotic. In the dishpan we have seen transitions between regimes of symmetric flow, steadily progressing waves, vacillation, and chaos, but I know of no cases in which the flow has assumed an extremely wide variety of patterns during an extended interval before regularly repeating itself. This strongly suggests, although it provides no proof, that if the atmosphere were not in a chaotic regime it would undergo rather simple periodic oscillations not appreciably more complicated than vacillation, with a period of perhaps a few weeks, although the quasi-biennial oscillation, if it could still exist in a nonchaotic regime, could upset things. Any simple behavior would also have to be modulated by the advance of the seasons, so that true repetition would occur only after a year, but each year could be a repetition of all of the previous ones.

Large migratory storms, which are features of both the dishpan and the atmosphere, would undoubtedly be found in our make-believe periodic atmosphere, and, because of the seasonal modulation, successive storms would travel on somewhat different paths. In the course of a year, a considerable portion of the earth's surface might then receive abundant rain, sufficient for agriculture, falling at each location on a particular set of dates. Without the seasons, rainfall would perhaps be confined to a few narrow belts.

At the other extreme in scale, thunderstorms and showers should be abundant enough to strike much of the earth in the course of a year. Hurricanes would be another matter, if they still occurred with a frequency characteristic of the real atmosphere. This could well be the case if the ocean-surface temperatures were comparable to those of the real oceans, since the formation and maintenance of hurricanes is strongly influenced by the temperature of the oceans beneath them. World-wide, a few dozen hurricanes might form during any year. Each might acquire a name, and the same name might be given to the identical hurricane arriving a year later, since the storm would be perceived as an annual event, just as El Niño, the sporadic warm current off the South American coast, is called El Niño whenever it returns. Thus there might have been a Hurricane Amy 1964, or a Hurricane Ben 1977.

Since every named hurricane would be following a track that hurricanes had been following for countless years, there would be little reason to expect much damage. Builders could avoid the paths of the stronger hurricanes, which together would not occupy too much real estate, but it would not be surprising if they built there anyway, presumably taking into account the known maximum wind speeds and the depths of any flash floods. In many respects, planning ahead without the vicissitudes of chaotic behavior would be a much simpler process. The greater difficulty in planning things in the real world, and the occasional disastrous effects of hurricanes and other storms, must therefore be attributed to chaos.

Weather forecasters using twentieth-century methods would not be needed, since this year's weather would be last year's. Meteorologists would still be active, just as tidal theorists are active in the real world, and they would seek explanations for the phenomena that they would be observing with such monotonous regularity. With the global circulation models that might be the fruits of a Global Atmospheric Research Program, modelers might succeed in simulating the significant weather structures, but it is doubtful that such a program would ever be initiated; with no need to improve the process of weather forecasting, who would supply the funds?

CHAPTER 4

Encounters with Chaos

Prologue

ONE OF THE TRIUMPHS of nineteenth-century computational mathematics was the discovery of Neptune. If you enjoy watching the early evening sky as one by one the planets announce their presence, you know that Neptune never takes part in the performance. It is far too faint, and, not surprisingly, it was first spotted in a telescopic search, but you may wonder how computation entered the picture. It is hard to say where the story begins, since every occurrence seems to have its antecedent, but a big event was the publication of Isaac Newton's *Principia Mathematica* in 1686. Once Newton's remarkable findings, including his laws of motion and his famous law of gravitation, had become generally recognized and accepted, astronomers found it fairly easy to write down systems of equations whose solutions would describe the motions of the planets in their orbits. These systems were special cases of the mathematicians' many-body problem, in which each of a number of objects is acted upon by the gravitational tug of all of the others. Solving the equations proved to be another matter.

Mathematicians typically do not feel that they have completely solved a system of differential equations until they have written down a general solution—a set of formulas giving the value of each variable at every time, in terms of the supposedly known values at some initial time. Consider, for example, a baseball just hit by a batter. The solution of the equations that govern its motion is easy to find. The formulas contain a symbol, say t, representing time; to find the ball's position and velocity at any moment it is sufficient to plug the appropriate value of t into the formulas and perform the indicated arithmetic operations. For the times when the ball is still in the air, the set of positions will form a parabola.

A still more general solution will take care of any baseball hit by any batter, whether fair or foul, whether a home run or an infield fly. The formulas will contain additional symbols, representing the position and velocity of the ball just when it leaves the bat.

Of course all this is what would happen if the air offered no resistance. Any golfer who has sliced a drive into the woods knows that golf balls do not travel in parabolas, and, to a lesser extent, the same thing is true of baseballs. Indeed, even the earth's curvature and rotation will prevent a path from being exactly parabolic. Equations are only models, and model baseballs and golf balls *can* travel along parabolas.

When we have had no luck in finding the general solution for some system, we can turn to numerical procedures. We may, for example, start with initial values of the variables, and, since differential equations are really formulas that tell us, in terms of the present values of all of the variables, how rapidly the values are changing, we can advance the solution forward in small time steps, until we reach the desired time. The procedure has the great advantage that it will often yield excellent approximations when other methods will yield nothing at all. It has the disadvantage that whenever we want a solution with new initial conditions or new constants, we must perform the computations all over again. Pure mathematicians have traditionally tended to hold numerical methods in scorn, and this attitude extended to the general use of computers in the days when computers were fairly new.

Eighteenth-century astronomers had little difficulty in finding the general solution for the two-body problem. The bodies move in elliptical orbits, which have a common focus at the combined center of mass. When the two bodies are the sun and a planet, the sun is so much more massive that the focus lies below the sun's surface, and the planet does most of the moving.

When theoretical solutions were compared with observations, say of Jupiter, there was good agreement, but the discrepancies, even though small, were too large to be attributed to errors in observation. An obvious suggestion was that they might be produced by the gravitational influence of the other planets, notably Saturn in Jupiter's case. Thus the three-body problem entered the picture. Here the early attempts to find a general solution all failed. In due time astronomers developed the perturbation method, which enabled them to use the two-body solution for the sun and one planet as a first approximation, and then to introduce the perturbing influence of a second planet as a correction. Perturbations by several planets could also be handled in this manner.

For most of the planets the method virtually removed the difference between observation and theory, but in the case of Uranus there was still a discrepancy that exceeded the expected observational error, which by the nineteenth century had become quite small. Gradually astronomers turned toward the idea that it could result from the gravitational pull of an as-yet-undiscovered planet.

Concurrently and independently, John C. Adams in England and Urbain J. J. LeVerrier in France undertook the formidable task of inverting the perturbation method; instead of computing the perturbation produced by a known planet moving in a known orbit, they asked how large a planet in what orbit could produce the known perturbation. In 1846, after many months of work and within a few days of one another, they came up with current positions only a few degrees apart. At LeVerrier's instigation a search was made at the Berlin Observatory, and it revealed the planet about one degree from his predicted position.

Such computational successes did not prevent mathematicians and astronomers from continuing to seek a general solution for the three-body problem. By analogy with the two-body problem it seemed reasonable that the solution was there, waiting to be uncovered. Chaotic behavior, which cannot usually be described by formulas into which values of time can be substituted, was not a part of the mathematics that they had learned, or in some instances had created.

What I want to do in this chapter is to present a few scenes from the drama of our growing awareness of chaos, from the time of the discovery of Neptune, when there was virtually no awareness at all, to a time nearly a century and a half later, when it was becoming apparent that chaos had been lurking almost everywhere. Of course I am not suggesting that the idea that minuscule events can lead to major consequences in everyday life or world affairs is something recent. The familiar bit of verse that begins, "For want of a nail, the shoe was lost," and progresses from the shoe to the horse to the rider to the battle to the kingdom, is not a twentieth-century creation. What does not seem to have been suspected in the middle nineteenth century is that phenomena governed by relatively uncomplicated laws, often expressible as deterministic mathematical equations, need not behave in the predictable manner that the laws or equations might lead one to suppose.

I shall make no attempt to tell a complete or unbroken story—neither space nor the extent of my knowledge of some of the significant events will allow this—nor shall I try to identify all of the important contributors or their specific contributions. Instead I shall present a rather personal and partly autobiographical account, offering my own analysis of some of the earlier developments, and discussing some of the more recent ones in terms of how I initially perceived them or how they subsequently influenced me. I shall devote considerable space—more than would belong in a balanced account—to the circumstances leading to my own findings. I hope nevertheless to leave my readers with a reasonably well balanced view of what took place to transform chaos from a hardly recognized phenomenon into one regarded as virtually ubiquitous.

Recognition

About thirty years after Neptune's discovery, when the three-body problem still seemed no closer to being solved, the American astronomer and mathematician George William Hill looked at a highly specialized case. Hill had already developed new approaches to the determination of planetary orbits, and he would back these up with extensive computations, often to fifteen decimal places. He now introduced three simplifications: first, one of the three bodies was assumed to have a negligibly small mass, so that the larger two were not influenced by it, and therefore satisfied the solvable equations for the two-body problem; second, the larger bodies moved about a common center in circles rather than more general ellipses; and third and perhaps most important, all three bodies moved in a single plane. In this way he reduced the problem to a system of four equations, with the four variables representing the small body's position and velocity in the plane. The equations were quite simple in appearance, but they still defied attempts to find a general solution.

With today's computers it is easy to determine particular solutions. Figure 35 shows a pair of possible orbits for the small body, which we may call a satellite, starting from nearby points with equal velocities. The coordinate system in which they are displayed rotates with the larger bodies, which we may call planets. That is, one of the coordinate axes, say the x-axis, is always parallel to the line joining the planets,

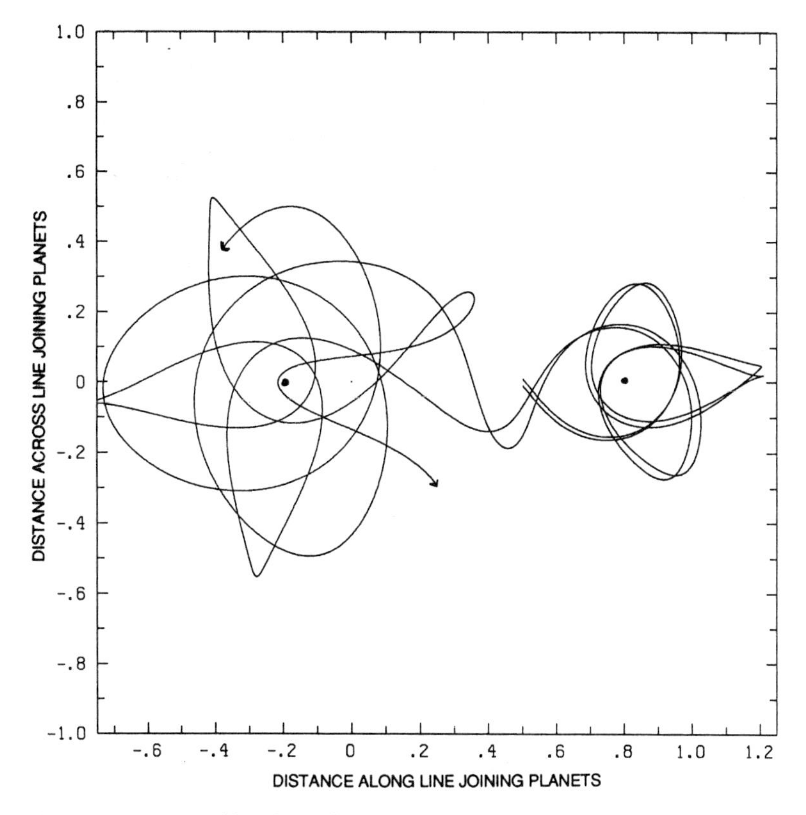

Figure 35. Two possible orbits of a satellite, starting with nearly identical conditions, as given by numerical solutions of Hill's reduced equations, extending for two years. The frame of reference from which the satellite is viewed rotates so as to make the planets, which are located 0.2 units to the left and 0.8 units to the right of the origin, and which are indicated by the dots, appear stationary.

while the other axis is always perpendicular, so that in the figure the planets occupy fixed positions, on the x-axis.

In this example, the planet on the left has four times the mass of the one on the right. The orbits first loop several times about the smaller planet, remaining rather close together, and then switch over to the larger one, diverging as they do so. After two years, when they reach the points indicated by the arrowheads, they are widely separated. Here a "year" means the time needed for the planets to make one revolution about their center of mass, which is one-fifth of the way from the larger to the smaller planet. A continuation would show the two orbits shut-

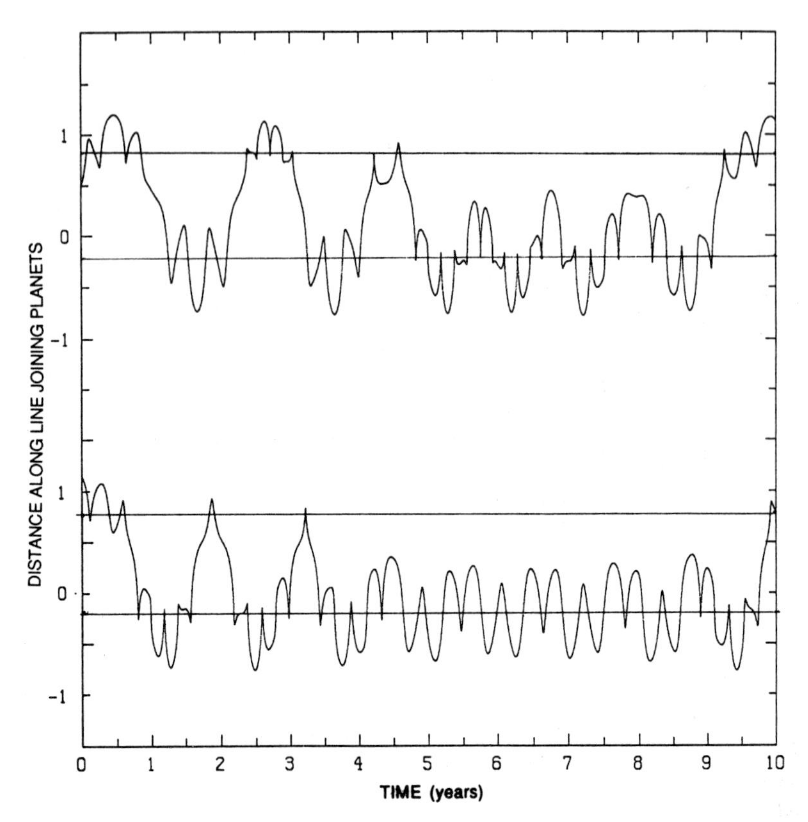

Figure 36. One of the orbits of Figure 35, extended for twenty years. The vertical coordinate here is the horizontal coordinate of Figure 35. The upper curve shows the first ten years, as indicated by the scale at the base, and the lower curve shows the next ten. The horizontal lines indicate the positions of the planets.

tling between the planets, and at distant future times they would be as likely to be looping around different planets as around the same one. Clearly the behavior is chaotic, but, as in many other Hamiltonian systems, other choices of initial states would have led to regular behavior, with orbits looping periodically about one planet or the other. Figure 36 is a graph of x, the distance of the satellite to the left or right of the center of mass, for one of the chaotic orbits, extended to twenty years. The irregular shuttling between the planets is apparent.

It is hard to imagine that Hill, who performed so many orbital computations, could not have produced something like Figure 35 in a few months or less, if he had wished to. He knew enough about the equations to have chosen, on his first try, an initial state that would force the satellite to shuttle between the planets, without escaping to infinity. What he presumably did not know was that, in a single orbit, the successive loops made about one planet or the other would permute their shapes in an irregular sequence. In any event, he was more interested in real cases, when the three bodies were the sun, earth, and moon, or Saturn, Titan, and Hyperion, and he knew that the moon could not shuttle between the earth and the sun.

Hill's system of equations has by now been solved many times—numerically. Evidently the solutions that produce the curves in Figure 35 are fairly typical; in *Does God Play Dice?*, Ian Stewart presents a rather similar curve, for a case where the two planets have equal masses.

The three-body problem, and in particular Hill's reduced problem, soon captured the imagination of the great French mathematician Henri Poincaré, born a few years after Neptune's discovery. Like others before him he failed to solve the *equations*, but unlike others he solved the *problem* in a very real sense; he proved that the equations could not be solved. Of course the equations do possess a general solution, but not one that will allow us to find it.

Poincaré did not obtain this amazing result overnight, and, in fact, he devoted many years to the task. Having found that a quantitative solution eluded him, he departed from the paths of previous investigators by turning to qualitative methods. He considered more general systems of equations than those of the reduced three-body problem, and in developing their properties he established the beginnings of a theory of dynamical systems.

To treat a system of n equations—for Hill's reduced problem, n is 4—he began by introducing phase space. This is the hypothetical n-dimensional space in which each state of the system is represented by a point, and particular solutions appear as special curves—solution curves, today generally called orbits. He then introduced the concept of a surface of section, today called a Poincaré section—an $(n - 1)$-dimensional manifold embedded in phase space and intersecting the solution curves. A simple example would be the set of points at which one of the n variables, say the first, assumes a particular value, say zero.

As we noted in the second chapter, a point where a solution curve intersects a surface of section completely determines the remainder of the curve, including the point where the curve next intersects the surface.

Thus, instead of studying the properties of entire curves, one may concentrate on sequences of intersections. A surface of section has two sides, which we might think of as being colored red and blue, and often only those intersections at which a curve crosses in one direction, say from the red side to the blue, are noted, while crossings from blue to red are disregarded. A simple periodic orbit, like a circle or an ellipse, will then cross the section at just one point—a fixed point—while more complicated periodic orbits may cross at several points before returning to the first one and repeating the cycle.

Poincaré next noted the possibility of solutions that he termed *asymptotic;* an asymptotic solution curve is one that approaches some periodic solution curve more and more closely as time advances, so that its sequence of intersections with a surface of section converges upon a fixed point. Other solutions can be asymptotic if the direction of time is reversed, that is, their sequences of intersections can appear to emanate from fixed points. Finally, there may be *doubly asymptotic* solutions, which are asymptotic in both directions of time. A sequence that emanates from a fixed point and subsequently converges to the *same* fixed point is called *homoclinic,* as is the fixed point itself. Poincaré demonstrated that the presence of a homoclinic point implies the existence of an infinite number of periodic sequences, with different periods, and also an infinite number of sequences that are not periodic. What he discovered through qualitative mathematical reasoning was chaos, at least in the limited sense.

Did he recognize the phenomenon of full chaos, where most solutions—not just special ones—are sensitively dependent and lack periodicity? He does not appear to have described his nonperiodic solutions as being sensitively dependent, but he was quite aware of the general phenomenon of sensitive dependence, and one of his most frequently quoted sentences begins, *"La prédiction devient impossible . . ."*

The quotation comes from one of his many philosophical writings—an essay on chance. Anyone unaware of Poincaré's work who might encounter the essay would at once recognize him as a deep thinker and a gifted writer, but might not realize that he was a mathematician, let alone perhaps the greatest mathematician of his day. In the essay he raises the possibility that what we generally regard as chance, or randomness, may in many instances be something that has of necessity followed from some earlier condition, even though we may be unaware that it has done so. He notes that in some cases we might be completely unable to detect the relevant antecedent condition, while in others we might observe it fairly accurately, but not perfectly. In the latter case the uncertainty might amplify and eventually become dominant. Is he not describing chaotic behavior?

The answer is not immediately clear. After some introductory discussion he analyzes four examples. The first is the general phenomenon of unstable equilibrium, which he illustrates by considering a cone standing on its vertex. In theory there is a position in which the cone can remain standing, but, as with the standing pencil, or the top that spins too slowly, the minutest disturbance will cause it to fall. If the disturbance is too small to observe, we cannot say in advance in which direction the cone will fall. It is here that he makes his statement that prediction becomes impossible.

Evidently he is not describing full chaos at this point. During its fall the cone is in a transient state. After transient effects have died out, the cone will be lying at rest on its side, in a state of stable equilibrium, and will exhibit no further seemingly random behavior. It will logically be predicted to remain in place, and any new small disturbances will not seriously upset the prediction.

His second example is the weather, and here he states, "We see that great disturbances are generally produced in regions where the atmosphere is in unstable equilibrium. Meteorologists know very well that the equilibrium is unstable, that a cyclone is going to be born somewhere; but where they are powerless to say." He does not seem to have gone deeply enough into meteorological theory to establish his statement, and presumably he is summarizing the ideas of the leading meteorologists of his day. Our ideas today are somewhat different—we recognize plenty of real instability, but few states of equilibrium except in models—but whether or not his assumptions were meteorologically correct has no bearing on whether the phenomenon as he visualized it constituted chaos. It does have a bearing on how we can go about determining whether he was visualizing chaos; if his atmosphere differed from ours, we cannot be guided by the realization that ours behaves chaotically.

Is the uncertainty that develops as a cyclone breaks out supposed to persist after the cyclone subsides, so that a prediction made before an outbreak, for the weather any time after the outbreak, will tend to fail, or is the uncertainty supposed to die out with the storm as more settled weather returns, so that predictions will fail only for the weather occurring while the storm is in progress? He does not say, and his other two examples—the distribution of the asteroids along the zodiac and the vicissitudes of roulette—do not reveal the answer.

However, he is only being deliberate. In a later section he describes a second phenomenon, which is equivalent to the persistence of uncertainty, and which he illustrates by considering the motions of individual molecules in a gas as they impinge on others. Here he states, "It suffices, we have just seen, to deflect the molecule before the collision by an infinitesimal, for it to be deflected after the collision by a finite quantity. If then the molecule undergoes two successive collisions, it will suffice to deflect it before the first collision by an infinitesimal of the second order, for it to be deflected after the first collision by an infinitesimal of the first order, and after the second collision, by a finite quantity."

The process is reminiscent of the pinball machine, except that each molecule strikes another moving molecule instead of a fixed pin. Poincaré does indeed recognize systems where the uncertainty will remain after the event that produces it has passed, and it is reasonable to conclude that he regarded the calm after the storm as being as unpredictable as the storm.

Nowhere in the essay does he mention the three-body problem, nor his novel way of treating it. Nevertheless, we are left with the feeling that he must have recognized the chaos that was inherent in the equations with which he worked so intimately.

In Limbo

It was in 1975 that Li and Yorke published their often-quoted paper "Period Three Implies Chaos." Whatever they may have intended to do, they succeeded in establishing a new scientific term, although one with a somewhat different meaning from what they had had in mind.

The same year comes as close as any to marking the onset of an "explosion" of scientific interest in chaos. The concurrence is presumably accidental. Like Juliet's rose, chaos by any other name would have smelled as sweet to most specialists. Some of them did not even find the name appropriate. In my own work I avoided it in favor of "irregularity" as recently as 1983. Soon after the burst of scientific interest there was a burst of general public interest, and here it seems highly probable that the catchy name was a decisive factor.

Since Poincaré had laid the foundations for dynamical-systems theory back in the nineteenth century, and had demonstrated that some systems behaved chaotically, at least in the limited sense, we may wonder why the explosion waited until some sixty years after his untimely death in 1912. Any explanation that I might offer must be speculative, but I believe that two factors were involved.

One of these was a general attitude toward irregularity. Poincaré was not seeking chaos. He sought to understand the orbits of the heavenly bodies, and he found chaos. To him it was the phenomenon that rendered the three-body equations too complex to be solved, rather than the principal subject of a future field of investigation.

Poincaré's successor was unequivocally George David Birkhoff, who wrote a definitive monograph entitled *Dynamical Systems* about fifteen years after Poincaré's death. Birkhoff had the distinction of being the first outstanding American mathematician to receive all of his formal education in the United States, and, by proving by example that it could be done, he gave a great boost to American mathematics. Even so, some of his colleagues used to say that his real teacher was Poincaré.

Birkhoff dealt with very general systems of equations. He produced rigorous proofs of some of Poincaré's conjectures. He discussed in detail the concept of a set of central solutions, which for the Hamiltonian equations of celestial mechanics becomes the set of all solutions, but which for many familiar dissipative systems becomes the attractor. By and large he gave the periodic solutions top billing.

Still, he was not unreceptive to irregularity. Like Poincaré, he had defined dynamical systems as systems governed by differential equations—flows—but, as we have seen, flows reduce to mappings when the whole of phase space is replaced by one of Poincaré's surfaces of section. He was thus inevitably led to the study of mappings, and, in a subsequent paper, he looked at a certain class of two-dimensional mappings in which an attracting set would contain a closed curve—one that forms a ring about an interior area and separates it from an exterior area. He noted that some mappings would continually carry certain points in one direction around the ring, while carrying other points in the opposite direction, and from this he deduced that the closed curve must double back on itself an infinite number of times. For some mappings the curves

Figure 37. The attracting set for the Poincaré mapping of the sled on the ski slope, when the moguls extend 1 meter above the pits; the set is also the attractor. The coordinate system is the one used in Figure 13.

were attractors, and strange ones at that. Perhaps they should not even be called curves, since in general one cannot go from one point of one of these curves to another point by traveling a finite distance along the curve.

To see one of the curves that Birkhoff knew about but never saw, we need look no farther than the ski slope. Sleds that move continually southeastward and those moving continually southwestward correspond to points in the attracting set that travel in opposite directions about a ring. Figures 37 and 38 give two examples. The first curve is the one that is being approached by the successive panels in Figure 11, where the moguls extend 1 meter above the pits; graphically it is indistinguishable from the attractor, shown in different coordinate systems in

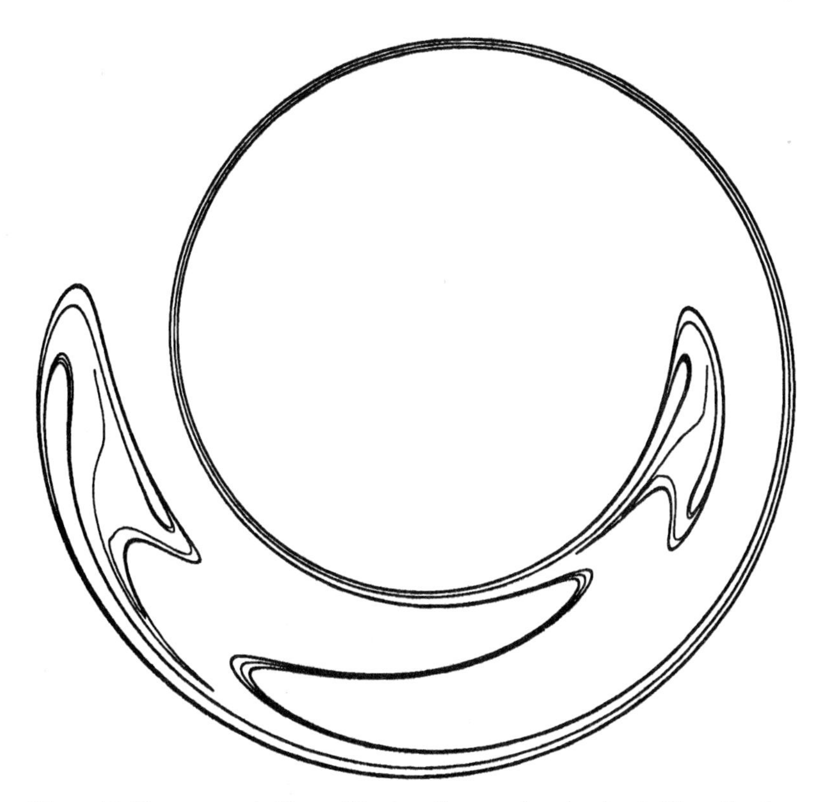

Figure 38. The same as in Figure 37, when the moguls extend only 50 centimeters above the pits; each of the two attractors consists of a single point of the set.

Figures 12 and 13. The second is the one approached by the panels in Figure 18, where the moguls are only half as high; here the attractor consists of only two points on the curve, and the chaos is limited.

Birkhoff did not offer specific sets of equations for his curves, but in 1945, the year following his death, Mary Cartwright and John Littlewood of Cambridge University studied the theoretical behavior of a periodically forced dissipative system—the so-called van der Pol oscillator—and they even presented their equation in their title, "On Non-linear Differential Equations of the Second Order. I. The Equation $\ddot{y}$-$k(1$-$y^2)\dot{y} + y = b\lambda\cos(\lambda t + \alpha)$, k Large." They found that under some conditions the system possessed two stable periodic solutions with different periods, and on this basis they stated that one of their invariant curves was like Birkhoff's "bad curve." Such was the attitude that prevailed to-

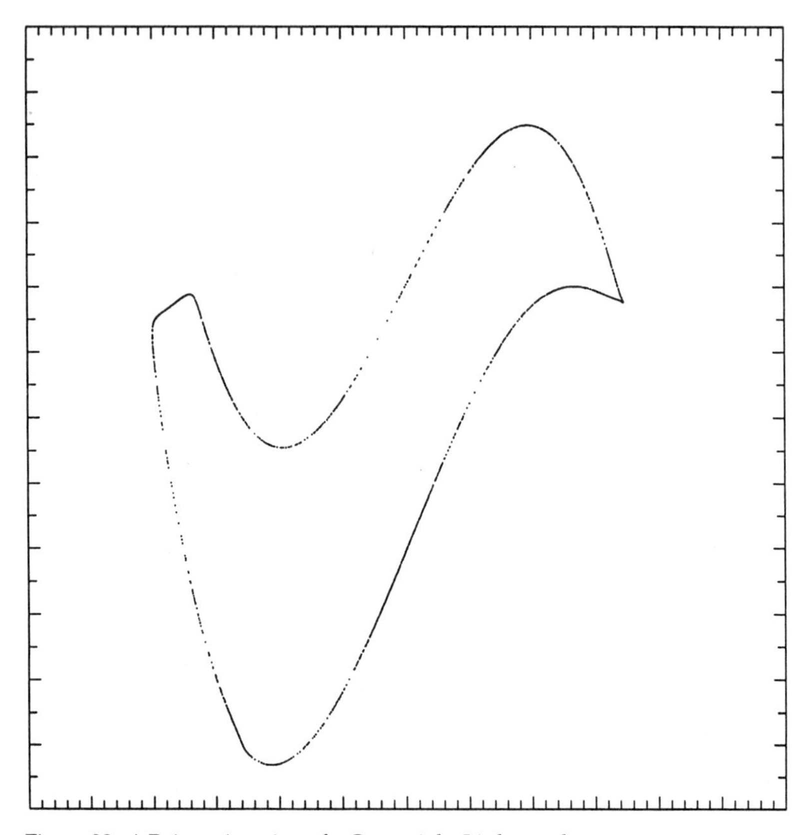

Figure 39. A Poincaré section of a Cartwright-Littlewood attractor.

ward the curves whose existence Birkhoff had deduced, even though he himself had only called them "remarkable."

Nevertheless, they did not abandon their curve, and they succeeded in identifying the first differential equation known to lead to a "bad" attracting set, and hence, except for some Hamiltonian systems without attracting sets, the first to produce at least limited chaos. The constants of the system are displayed in the title of the paper, and Figure 39 shows a Poincaré section of the attractor, for a rather rare choice of values of the constants that leads to full chaos; thus the figure shows a strange attractor. As such it is not spectacular; the ubiquitous sets of parallel segments are so closely packed that even a substantial enlargement will not resolve them. Only the presence of sensitive dependence reveals that the attractor must be strange.

One may argue that the absence of an early outburst was not *caused* by a prevailing lack of interest; it *was* the lack of interest. To some extent this is true, yet it may have been caused by the priorities of the leaders in the field. One of the quickest ways for a young scientist to gain recognition, and perhaps a prize, is to solve a problem that has become well known because the leading scientists of an earlier generation have tackled it and failed. One who is seeking such recognition may have little incentive to start out in a totally new direction, even though history indicates that the vast unexplored territory surrounding new problems sometimes holds the key to the solution of older ones. Certainly Poincaré and Birkhoff and most other leaders did not suggest that the problems of the future would lie in chaos theory.

Still, chaos could not remain in limbo forever. A decidedly more positive attitude began to appear as the fifties advanced and gave way to the sixties. It is clearly evident, for example, in the work of the American mathematician Stephen Smale, creator of the famed horseshoe.

The horseshoe is a two-dimensional mapping. To formulate it, imagine that you take a square in phase space, compress it vertically and stretch it horizontally, and then bend it into a shape resembling a horseshoe and lay it over its original position. You may do this in many ways, but the extremities of the horseshoe should project from the sides of the square that were compressed, not those that were stretched. Each point of the original square will then be moved to a point in the horseshoe, but not necessarily to a point in the square itself. The points that will remain within the square forever as the procedure is iterated, forward or backward, form an invariant set, and Smale showed that within this set, regardless of the particular shape of the horseshoe, there is chaos; most sequences of iterates are nonperiodic. Hence, within the whole system, there is at least limited chaos. Whether or not the invariant set is an attractor, and hence whether or not there is full chaos, depends on what happens to the points after they leave the square, and in particular on whether they reenter after further iteration.

Sometimes a horseshoe can be discovered within a mapping that in its totality may not be suggestive of a horseshoe. This is the situation illustrated in Figure 40. The mapping is accomplished by taking the square, giving it a quarter turn counterclockwise, compressing it a bit vertically,

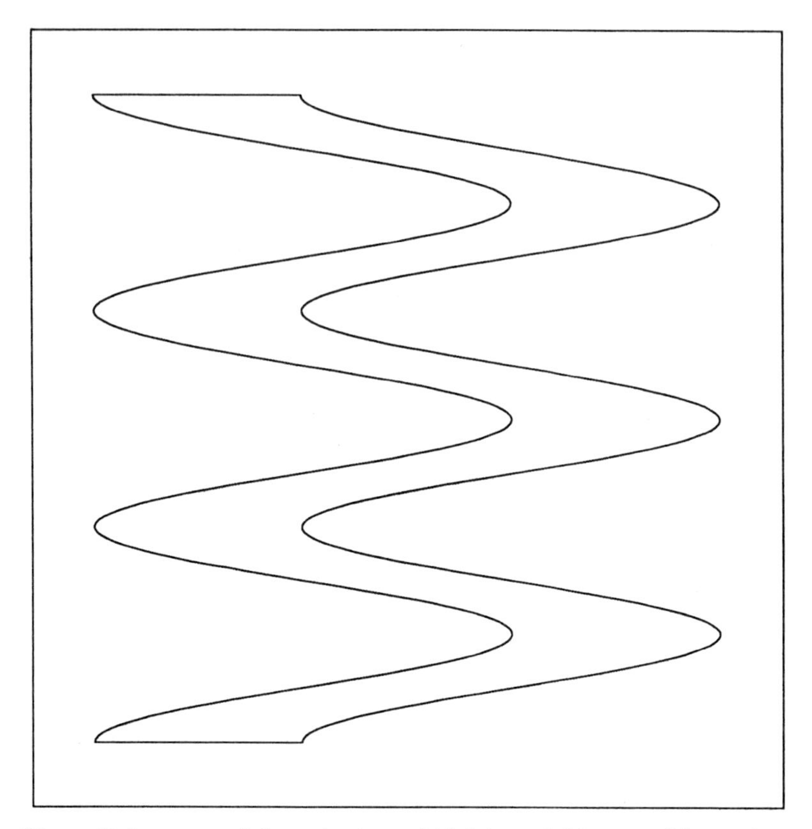

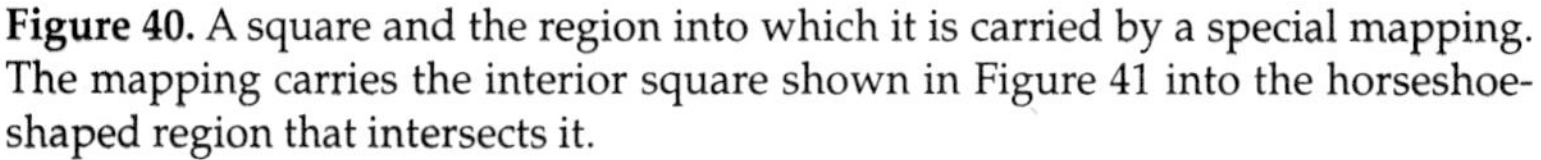

Figure 40. A square and the region into which it is carried by a special mapping. The mapping carries the interior square shown in Figure 41 into the horseshoe-shaped region that intersects it.

and then squeezing it horizontally so that it fits into the contorted region inside. This mapping carries the small interior square shown in Figure 41 into the horseshoe that intersects it.

Whatever the nature of the chaos, Smale was interested in the "bad" curves as well as the "good" ones. More generally, what appeared to be emerging was a clear interest in the possibility of systems with at least *some* irregular solutions, often without much concern for whether the general solutions were irregular. To a mathematician, limited chaos can be as fascinating as full chaos; it leads to the same peculiarly shaped figures—observe Figure 38—even though they may not be attractors, and for specific systems its existence can often be established when a proof of full chaos remains elusive. Yet to someone interested in tangible systems

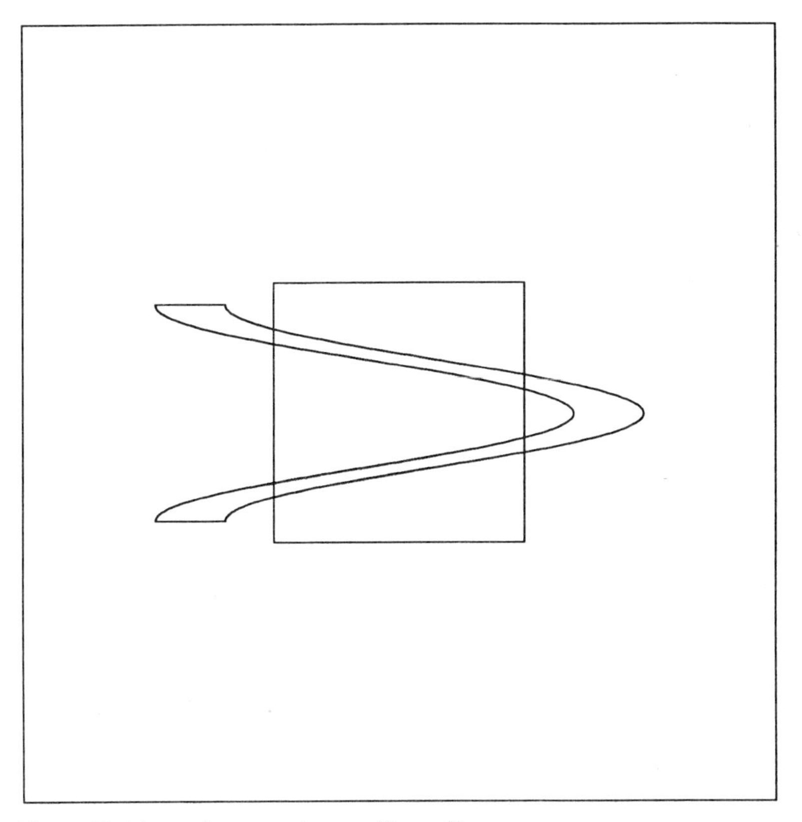

Figure 41. A horseshoe mapping; see Figure 40.

there is a vast difference between limited and full chaos, since the former will not be observed. If potential chaos in the heartbeat were no more than limited, we would not have to worry about arrhythmias. If the global weather system possessed only limited chaos, the weather would be as predictable as if there were no chaos at all.

Several years later Smale wrote a paper called "Differentiable Dynamical Systems," which has become the acknowledged successor to Birkhoff's treatise. He immediately departs from tradition by defining dynamical systems in terms of *difference* equations—mappings—and he notes the greater simplicity that follows. He includes a discussion of the horseshoe, and presents a now-familiar three-dimensional mapping often called the solenoid map. Whether or not the strange set produced by the horseshoe is part of an attractor, the solenoid unequivocally possesses a strange attractor. It resembles an infinite wire as it might be coiled inside a tube, but a cross section is unlike anything that we have seen so far. No two points are connected, although in the sample in Figure 42 some of them may seem connected, since they are closer together than the diameters of the printed dots.

Whatever influence the early attitudes toward irregularity may have had, the bigger factor in the timing of the outburst of interest in chaos was surely the arrival of the computer. The works of such people as Birkhoff, in the precomputer age, and Smale, when computers were commonplace, show clearly that a theory of some aspects of chaos can be fully developed with pencil and paper. Demonstrating chaotic solutions of specific systems of equations, and constructing the accompanying strange attractors, require something more. Let me cite some of my own work as an illustration.

The first mathematically generated chaos that I encountered was produced by a very crude model of the global weather system, which contained not thousands or millions of variables, but just twelve. Solving a system of this size by hand computation is not an impossibly long task, as the astronomers of the eighteenth and nineteenth centuries aptly demonstrated with their calculations of planetary orbits. If, after formulating the model, I had chosen the values of the constants that I finally used, I might, by dropping all other professional activities, have computed a recognizably nonperiodic solution by hand in a month or two—a time comparable to what I would have needed afterward to write up the result for publication. Lack of a computer might have done little more than double the total effort. The point is that at first I did not know what values of the constants would lead to chaos, nor whether any such values even existed, and I had to make many tries before finding some that worked. Before settling on the twelve-variable model, I had experimented unsuccessfully with other systems of equations. Without the computer the needed time for computation alone would have been years instead of months, and, with other problems to occupy much of my time, I would probably not have continued. Even if I had been lucky on the first try, determining a large collection of solutions, and producing from them a set of illustrations like the ones in this volume, would have been out of the question. I feel certain that anyone else who might have sought nonperiodic solutions of differential equations by hand computation would have met with similar obstacles.

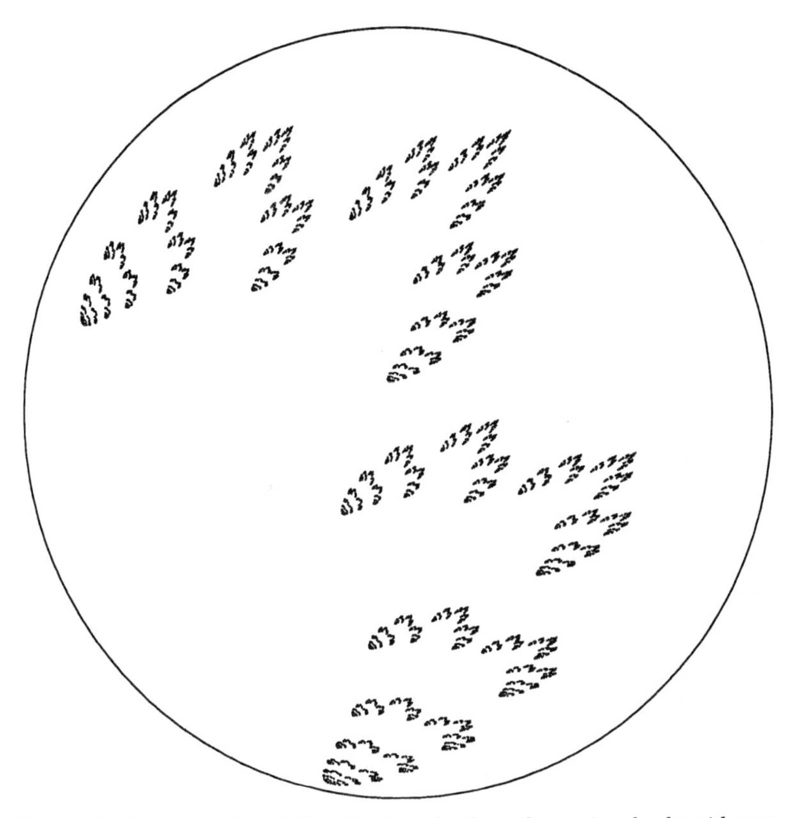

Figure 42. A cross section of the attractor of a three-dimensional solenoid mapping, with a circular cross section of the tube in which it is contained.

It seems reasonable to conclude that, once computers had been around for a while, their contribution to the growing awareness of chaos extended well beyond their application to the hitherto unsolvable equations that arose in specific problems pursued by individual scientists. Once a sufficient number of investigators had published their solutions in the open literature, other scientists were able to discern certain common features that had not been previously recognizable, and it no longer appeared that the study of similar equations for their own sake would prove fruitless. With the new incentive came new investigations of equations of all sorts. Suddenly, the strange attractors that had sometimes shown up in the earlier problems were joined by a host of others, often produced by equations that bore no obvious relation to specific physical problems, and were accepted as their own justification.

Searching

At this point I want to recount in considerable detail the circumstances that led to my personal involvement with chaos. I readily recall the main events, but my attempts to relate them to earlier and later developments are bound to involve some speculation. The setting was the Department of Meteorology at the Massachusetts Institute of Technology, once familiarly known as Boston Tech, but now almost invariably called M.I.T. I had been doing postdoctoral work there since 1948, and my main interest was the dynamics of atmospheric structures of global and continental size.

As a student I had been taught that the dynamic equations determine what takes place in the atmosphere. However, as my thinking became more and more influenced by numerical weather forecasting, it became evident to me that these equations do not prohibit *any* atmospheric state, realistic or unrealistic, from being an *initial* state in a solution. It must be, I felt, that the various solutions of the equations all converge toward a special set of states—the realistic ones. I had even made a few unsuccessful attempts to find formulas for this set, and had already abandoned the effort. In the light of today's knowledge it appears that I was seeking the attractor, and was right in believing that it existed but wrong in having supposed that it could be described by a few formulas.

The opening scene took place in 1955, when Thomas Malone resigned from our faculty in order to establish and head a new weather research center at the Travelers Insurance Company in Hartford, Connecticut. Tom had been directing a project in statistical weather forecasting, a field that had gained a fair number of adherents in the early days of computers.

Philosophically, statistical forecasting is more like synoptic than dynamic forecasting, in that it is based on observations of what has happened in the past, rather than on physical principles. It is like dynamic forecasting in that it makes use of values of the weather elements at particular locations, rather than identifiable synoptic structures. The type of statistical forecasting that had received most attention was "linear" forecasting, where, for example, tomorrow's temperature at New York might be predicted to be a constant *a*, plus another constant *b* times today's temperature at Chicago, plus another constant *c* times yesterday's relative humidity at St. Louis, plus other similar terms.

There were long-established mathematical procedures for estimating the optimum values of the constants *a, b, c,* etc., and, in fact, about the only opportunity for the meteorologist to use any knowledge of the atmosphere was in selecting the predictors—the weather elements to be multiplied by the constants. The computational effort that goes into establishing a formula increases rapidly with the number of predictors, and, as with numerical weather prediction, the work proliferated only after computers became reasonably accessible. The method was regarded by many dynamic meteorologists, particularly those who were championing numerical weather prediction, as a pedestrian approach that yielded no new understanding of why the atmosphere behaved as it did.

I was appointed to fill the vacancy that Tom's departure had created, and with his job I also acquired his project. During the next year I examined numerous statistically derived formulas, and finally convinced myself that what the statistical method was actually doing was attempting to duplicate, by numerical means, what the synoptic forecasters had been doing for many years—displacing each structure at a speed somewhere between its previous speed and its normal speed. One-day prognostic charts were decidedly mediocre, although the method was and still is useful for deciding what local weather conditions to predict, once a prognostic chart is available.

Needless to say, many of the devotees of statistical forecasting disagreed with my findings. Possibly they looked upon me as an infiltrator from the numerical weather prediction camp. In particular, some of them pointed to a recent paper by the eminent mathematician Norbert Wiener, which appeared to show that linear procedures could perform as well as any others, and so necessarily as well as numerical weather prediction or synoptic forecasting. I found this conclusion hard to accept, and convinced myself, although not some of the others, that Wiener's statements, which were certainly correct but were not written in the most easily understandable language, were being misinterpreted. At a meeting in Madison, Wisconsin, in 1956, attended by a large share of the statistical forecasting community, I proposed to test the hypothesis by selecting a system of equations that was decidedly not of the linear type. I would use a computer to generate an extended numerical solution, and then, treating the solution as if it had been a collection of real weather data, I would use standard procedures to determine a set of optimum linear prediction formulas. If these formulas could really match up to any other forecasting scheme, they would have to perform perfectly, since one could easily "predict" the "data" perfectly simply by running the computer program a second time.

My first task was to select a suitable system of equations. I proceeded in the manner of a professional meteorologist and an amateur mathematician. Although in principle a wide variety of systems would have worked, I was hoping to realize some side benefits by choosing a set of equations resembling the ones that describe the behavior of the atmosphere. After some experimentation I decided to work with a drastically simplified form of the filtered equations of numerical weather forecasting, which would reduce the number of variables from the many thousands generally used to a mere handful.

One day Robert White, a postdoctoral scientist in our department who later went on to become Chief of the United States Weather Bureau, and still later headed the organizations that superseded it, suggested that I acquire a small computer to use in my office. If you wonder why I had not already done so, recall that this was more than twenty years before personal computers first appeared on the market. In fact, computers for personal use were almost unheard of, and the idea had certainly not occurred to me. We spent several months considering various competing models and finally settled upon a Royal-McBee LGP–30, which was about the size of a large desk and made a continual noise. It had an internal memory of 4096 32-bit words, of which about a third had to be reserved for standard input and output programs. It performed a multiplication in 17 milliseconds and printed a full line of numbers in about 10 seconds. Even so, when programmed in optimized machine language it was about a thousand times as fast as a desk calculator—pocket calculators had not yet appeared—and was ideal for solving small systems of equations.

It should not surprise us that in a day when computers were far from ubiquitous, most scientists, myself included, had not learned to write computer programs. I spent the next few months getting acquainted with the computer. Upon returning to the simplified meteorological equations, I settled on a form with fourteen variables. Later I cut the number to thirteen and then to twelve by suppressing the variations of one and then two of the variables.

The equations contained several constants that specified the intensity and distribution of the external heating needed to drive the miniature atmosphere. Thus, if one set of constants failed to produce a useful solution, there were always others to try. My early attempts to generate "data" invariably produced "weather" that settled down to a steady state and was therefore useless for my purposes. After many experiments, I at last found a solution that unmistakably simulated the vacillation observed in the dishpan. I eagerly turned to the procedure for determining the best linear formula, only to realize that *perfect* linear prediction was possible simply by predicting that each variable would assume the value that it had assumed one vacillation cycle earlier. It was then that I recognized that for my test I would need a set of equations whose solutions were not periodic. What I did not even suspect at the time was that any such set would have to exhibit sensitive dependence.

By this time it was 1959. Although by now I had become a part of the statistical forecasting community, I managed to retain my status as a dynamicist, and I planned to attend a symposium in numerical weather prediction to be held the following year in Tokyo. Titles for the talks were due well in advance. I gambled on finding a suitable system of equations and completing my test, and submitted the title "The Statistical Prediction of Solutions of Dynamic Equations."

If I had been familiar then with Poincaré's work in celestial mechanics, it might have made sense for me to abandon the twelve equations and turn to the four equations of Hill's reduced problem, which, besides already being known to possess some nonperiodic solutions, were a good deal simpler. My guess, though, is that such a switch would not have appealed to me; the mere knowledge that simple systems with nonperiodic solutions did exist might have given me additional encouragement to continue my own search, and in any case I still had my eye on the possible side benefits. These, I felt, demanded that I work with a dissipative system. As it was, I kept trying new combinations of constants, and finally encountered the long-sought nonperiodic behavior after making the external heating vary with longitude as well as latitude. This is of course what happens in the real atmosphere, which, instead of receiving most of its heat directly from the sun, gets it from the underlying oceans and continents after they have been heated by the sun. Continents and oceans differ considerably in their capacity to absorb solar energy, and in the manner in which they subsequently transfer it to the atmosphere. When I applied the standard procedure to the new "data," the resulting linear forecasts were far from perfect, and I felt that my suspicions had been confirmed.

The solutions proved to be interesting in their own right. The numerical procedure advanced the weather in six-hour increments, and I had programmed the computer to print the time, plus the values of the twelve, thirteen, or fourteen variables, once a day, or every fourth step. Simulating a day required about one minute. To squeeze the numbers onto a single line I rounded them off to three decimal places, and did not print the decimal points. After accumulating many pages of numbers, I wrote an alternative output program that made the computer print one or two symbols on each line, their distances from the margin indicating the values of one or two chosen variables, and I would often draw a continuous curve through successive symbols to produce a graph. It was interesting to watch the graph extend itself, and we would sometimes gather around the computer and place small bets on what would happen next, just as meteorologists often bet on the next day's real weather. We soon learned some of the telltale signs for peculiar behavior; in effect, we were learning to be synoptic forecasters for the make-believe atmosphere.

In Figure 43 we see a copy of fifteen months of the somewhat faded original output, divided for display purposes into three five-month segments. The chosen variable is an approximate measure of the latitude of the strongest westerly winds; a high value indicates a low latitude. There is a succession of "episodes," in each of which the value rises abruptly, remains rather high for a month or so, and then drops equally abruptly, but the episodes are not identical and are not even equal in length, and the behavior is patently nonperiodic.

At one point I decided to repeat some of the computations in order to examine what was happening in greater detail. I stopped the computer, typed in a line of numbers that it had printed out a while earlier, and set it running again. I went down the hall for a cup of coffee and returned after about an hour, during which time the computer had simulated about two months of weather. The numbers being printed were nothing like the old ones. I immediately suspected a weak vacuum tube or some other computer trouble, which was not uncommon, but before calling for service I decided to see just where the mistake had occurred, know-

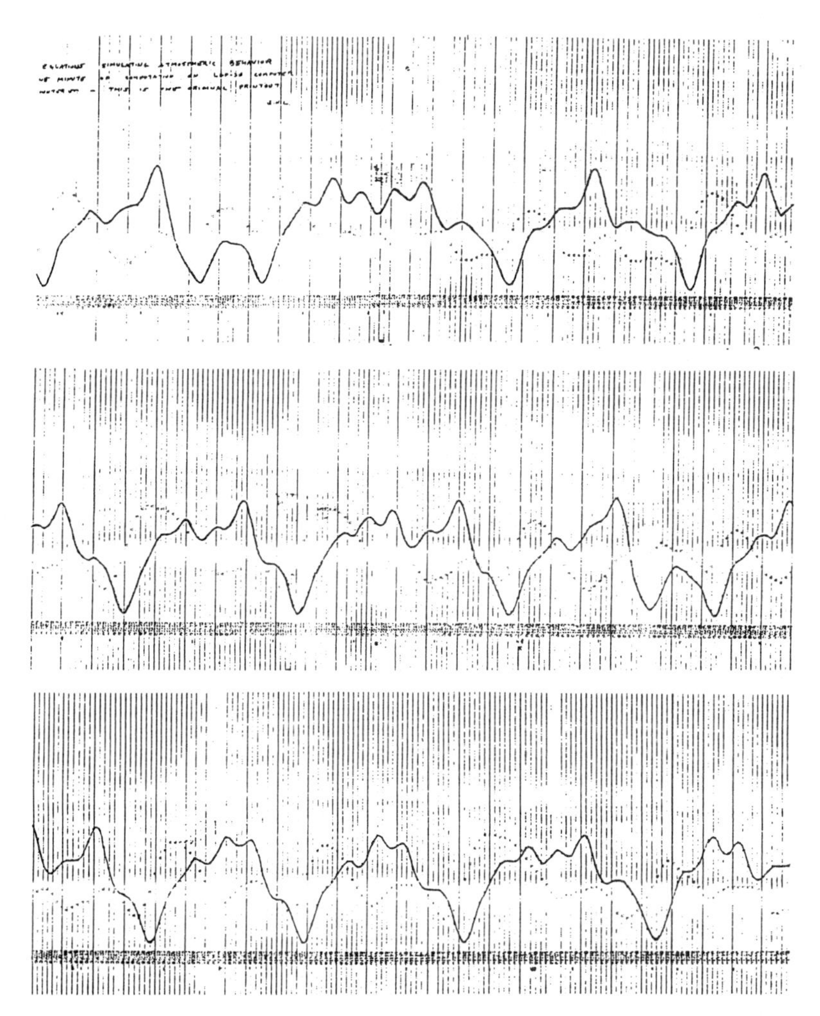

Figure 43. A fifteen-month section of the original print-out of symbols representing two variables of the twelve-variable model. A solid curve has been drawn through the symbols for one variable, while the symbols for the other are faintly visible. The section has been broken into three five-month segments, shown on consecutive rows.

ing that this could speed up the servicing process. Instead of a sudden break, I found that the new values at first repeated the old ones, but soon afterward differed by one and then several units in the last decimal place, and then began to differ in the next to the last place and then in the place before that. In fact, the differences more or less steadily doubled in size every four days or so, until all resemblance with the original output disappeared somewhere in the second month. This was enough to tell me what had happened: the numbers that I had typed in were not the exact original numbers, but were the rounded-off values that had appeared in the original printout. The initial round-off errors were the culprits; they were steadily amplifying until they dominated the solution. In today's terminology, there was chaos.

It soon struck me that, if the real atmosphere behaved like the simple model, long-range forecasting would be impossible. The temperatures, winds, and other quantities that enter our estimate of today's weather are certainly not measured accurately to three decimal places, and, even if they could be, the interpolations between observing sites would not have similar accuracy. I became rather excited, and lost little time in spreading the word to some of my colleagues.

In due time I convinced myself that the amplification of small differences was the *cause* of the lack of periodicity. Later, when I presented my results at the Tokyo meeting, I added a brief description of the unexpected response of the equations to the round-off errors.

The Strange Attractor

In 1971 I attended a turbulence meeting in La Jolla, California. Some of the same old turbulence people were there, and I more or less expected to hear some of the same old ideas, but there was a newcomer—the French mathematical physicist David Ruelle—whose scheduled talk was entitled "Strange Attractors as a Mathematical Explanation of Turbulence." The title seemed strange to me, and I even asked a colleague if it might be a mistranslation from the original French. He assured me that it was not, and when Ruelle spoke, in English at least as fluent as mine, I realized that even though I had not heard of a strange attractor, I had seen one. Let me describe the circumstances.

Ruelle's talk was a summary of a paper that he and Floris Takens had just published under the title "On the Nature of Turbulence," which first contained the expression "strange attractor," and which, like Smale's treatise on dynamical systems, became one of the most frequently cited papers in the field of chaos. In the paper they used the solenoid mapping—the one that produces Figure 42—as an illustrative example, and they described turbulent motion as being "chaotic."

At the Tokyo meeting more than a decade earlier I had briefly mentioned the unexpected behavior of the twelve-variable model, but I felt that a full discussion of the relationship between lack of periodicity and growth of small differences, and its implications for long-range weather forecasting, belonged in a separate paper. For that paper I was anxious to use an even simpler system of equations as a principal illustrative example, in the hopes of being able to demonstrate exactly what was happening. I tried to simplify the model still more without losing the sensitive dependence, but with no luck. Actually there was a way to reduce it to three variables, but I never discovered it until 1983.

My search came to an abrupt end one afternoon in 1961 when I was visiting Barry Saltzman at the Travelers Weather Center; this was the center that Tom Malone had established several years earlier. Barry showed me a system of seven equations that he had been solving numerically. The equations were a bit like mine, but they modeled convective fluid motion driven by heating from below, such as might occur locally over warm terrain, instead of the global atmospheric circulation, which is driven mainly by horizontal differences in heating. He was interested in periodic solutions and had obtained a number of them, but he showed me one solution that refused to settle down.

I looked at it eagerly, and noted that four of the seven variables soon became very small. This suggested that the other three were keeping each other going, so that a system with only these three variables might exhibit the same behavior. Barry gave me the go-ahead signal, and back at M.I.T. the next morning I put the three equations on the computer, and, sure enough, there was the same lack of periodicity that Barry had discovered. Here was the long-sought system whose existence I had begun to doubt.

I was lucky in more ways than one. An essential constant of the model is the Prandtl number —the ratio of the viscosity of the fluid to the thermal conductivity. Barry had chosen the value 10.0 as having the order of magnitude of the Prandtl number of water. As a meteorologist, he might well have chosen to model convection in air instead of water, in which case he would probably have used the value 1.0. With this value the solutions of the three equations would have been periodic, and I probably would never have seen any reason for extracting them from the original seven.

The three equations do not describe real convective motions very

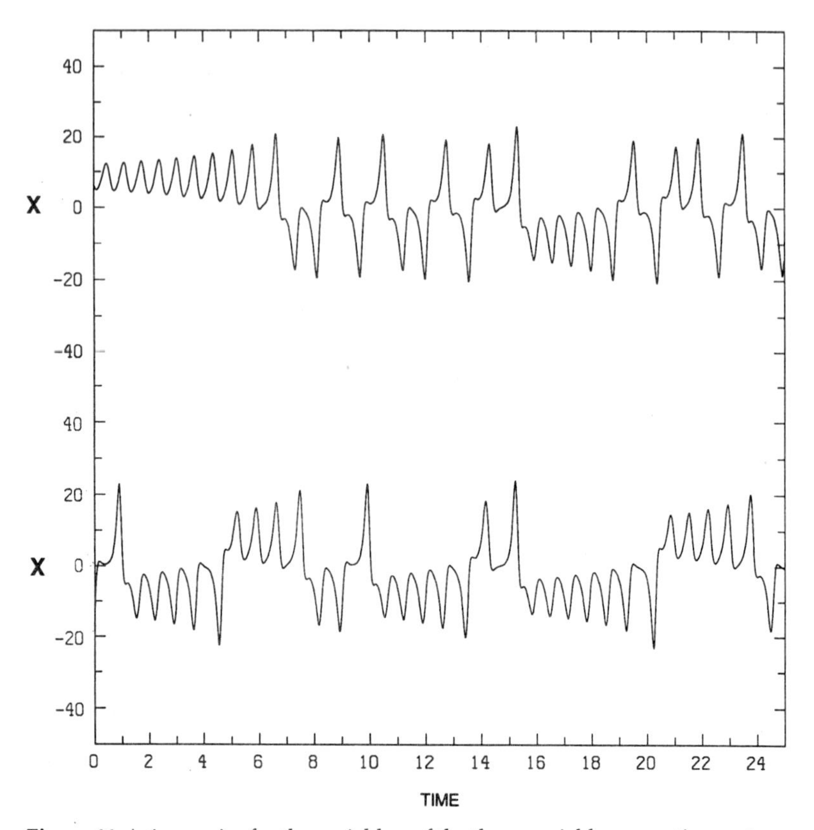

Figure 44. A time series for the variable x of the three-variable convective system. The series, extending for fifty time units, has been broken into two segments, shown on consecutive rows.

well, but for my purposes they did not need to. Treated together as a mathematical abstraction, they demonstrate one of the simplest ways in which a deterministic system can behave if it is not going to behave periodically. Figure 44 is a graph of one of the variables against time. Evidently the system undergoes amplifying oscillations about one state, which is actually a state of unstable equilibrium, until the oscillations become too strong. It then proceeds to oscillate about another unstable state, until these oscillations become too strong, after which it continually alternates between its two modes of behavior. The important feature, clearly revealed in the figure, is that the successive numbers of oscillations about one state or the other occur in an irregular sequence. Indeed, the behavior has something in common with the shuttling of the satellite between two planets that we saw in Figure 36, and, even though one can no more write a set of formulas for the general solution here than in the case of the satellite, the graphical solution reveals, for the chosen values of the constants, virtually everything that can happen. It is a nearly complete substitute for a general solution.

I decided to use the system as the illustrative example in the write-up. I felt at that time that the really important finding was the fact that under fairly general conditions a lack of periodicity implied limited predictability, rather than the discovery of a specific system of equations with nonperiodic solutions. I also felt, and still feel, that of the total systematic quantitative effort that has been put into predicting various phenomena, the biggest share has gone into weather forecasting; witness the world-wide network of observing stations and the development of five-million-variable models. In any event I addressed the paper primarily to meteorologists, and submitted it to the *Journal of the Atmospheric Sciences,* originally with the title "Deterministic Turbulence." Soon I changed it to "Deterministic Nonperiodic Flow," after the editor, whom I knew well—it was Norman Phillips, creator of the first global circulation model—persuaded me that the equations lacked some of the properties that we generally associate with turbulence.

I also managed to fulfill a long-standing wish to look at an attractor. I simply chose two of the variables as coordinates on a plane, and plotted numerical values of the third variable beside the points that I plotted, just as I had so often worked with longitude and latitude as coordinates on a weather map, and plotted values of pressure. I then drew contours of the third variable, just as I had often drawn isobars, except that the procedure was slightly more complicated, because there turned out to be two distinct sets of contours over part of the plane, implying that the attractor was composed of two distinct surfaces, one above the other. These appeared to merge as one followed an orbit in either surface toward the base of the figure.

Figure 45 is the complete map of the attractor, with its two merging sets of contours, while Figure 46 shows the outline of the attractor and an extended solution curve that lies within it. Viewed from a different angle, the curve would become the butterfly. One can see how, in three dimensions, the curve can pass continually from one surface to the other, without ever intersecting itself. I decided to include a description of the attractor in the write-up.

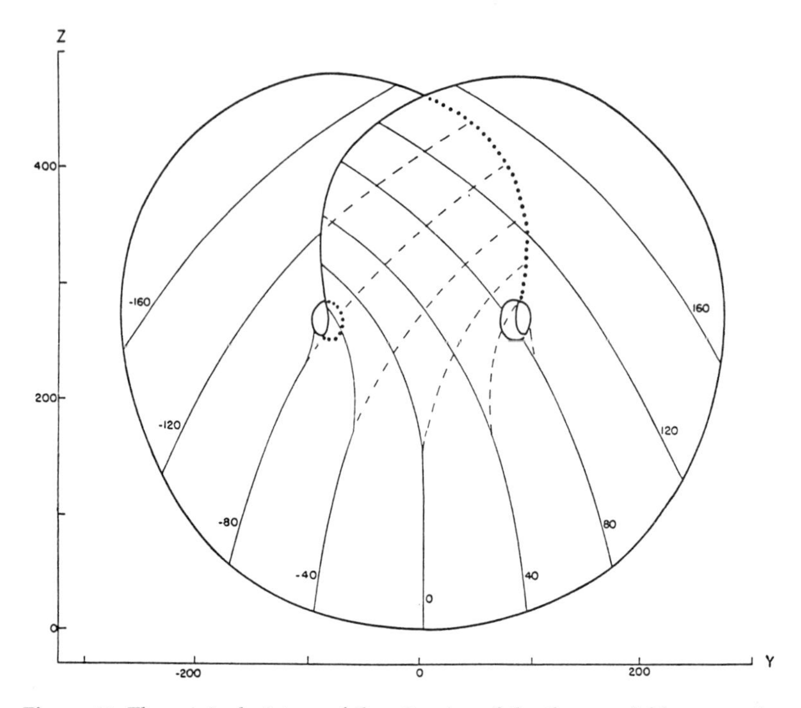

Figure 45. The original picture of the attractor of the three-variable convective system. With the variables y and z as coordinates, contours of the variable x have been drawn. Where there are two resolvably different values of x, the contours for the lower value are dotted. Figure reproduced by permission of the American Meteorological Society.

I have always told my students, when they are writing up some of their work, to allow much more time than they think they can possibly need—at least a month in the case of a thesis. In trying to write a clear explanation of some point that you think you understand, you may discover that you do not understand it well enough to explain it, and you may need to spend quite a while to think it through, perhaps performing some more pencil-and-paper research or even some computations. This advice applies to teachers as well as to students. In my case, in trying to understand and then explain to the reader how the two surfaces could merge, even though theory indicated that two solution curves—in this case one in each surface—could never merge, I realized that the surfaces—not just the curves—must fail to merge; if they did merge, there

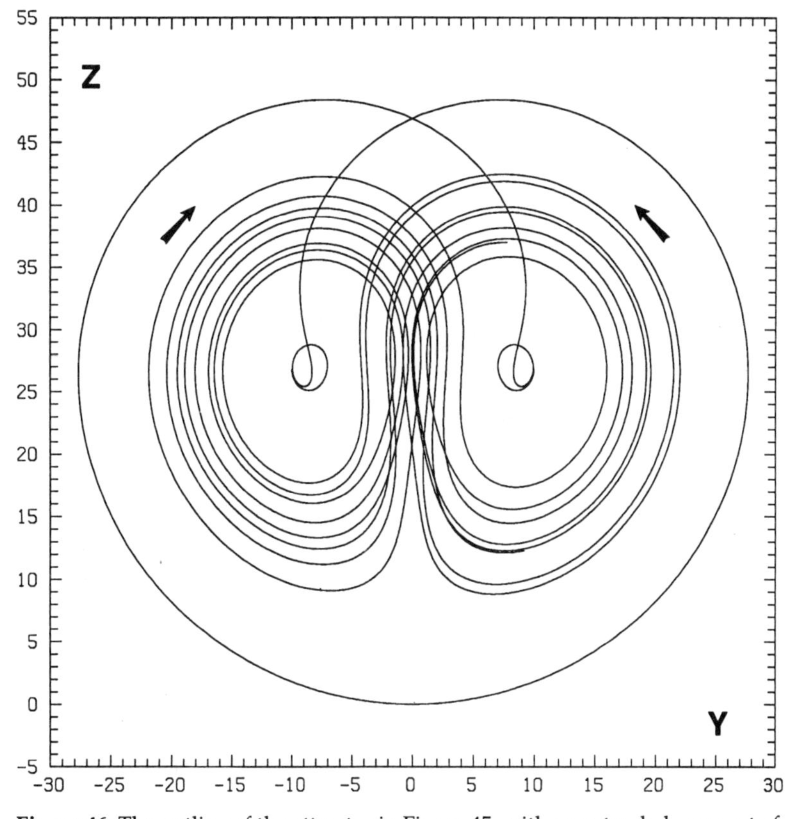

Figure 46. The outline of the attractor in Figure 45, with an extended segment of an orbit contained in the attractor. The orbit alternates irregularly between clockwise circuits about the left hole and counterclockwise circuits about the right.

would be one particular curve on one surface that would merge with any prechosen curve on the other, violating the theory. The only remaining possibility was that the surfaces merely appeared to merge, and actually maintained their identities as one followed a solution curve. Where a surface appeared to pass under another, then, there were two surfaces passing under two others, making a total of four, but these all had to maintain their identities, so that there would have to be four surfaces passing under four others, etc. My final conclusion, now thoroughly substantiated, was that the surfaces were really infinite in number; as we say now, the attractor was strange. I never guessed at the time that this attractor would for a while, much later, become the feature of the paper that would draw the most attention.

In preparing the paper I had been guided by Birkhoff's treatment of dynamical systems. When I submitted the paper, the reviewer, who proved to be quite conscientious and competent, noted that some of my results overlapped some of the works that had postdated Birkhoff's monograph, and he specifically mentioned the recently translated textbook *The Qualitative Theory of Differential Equations* by the mathematicians V. V. Nemytskii and V. V. Stepanov, originally published in Russian in 1946. Part II of the book is a complete treatment of dynamical systems, which, as one might guess from the title, are defined in terms of differential equations. In it the authors define a solution as being "stable in the sense of Lyapunov"—the Russian mathematician Aleksandr Lyapunov was a contemporary of Poincaré's and a pioneer in stability theory—if any other solution originating sufficiently close to it remains close to it as time progresses. This is simply the absence of sensitive dependence. They demonstrate that, under certain conditions, which the equations with which I was working evidently satisfied, a solution possessing Lyapunov stability must be periodic or almost periodic. They trace proofs of various forms of the result to Philip Franklin in 1929 and to A. A. Markov, son of the A. A. Markov of "Markov process" fame, in 1933.

It follows immediately that a nonperiodic solution is unstable in the sense of Lyapunov, or sensitively dependent, and one more step shows that if the general solution of a system is nonperiodic, the general behavior is chaotic. What I had thought was my principal result suddenly lost much of its perceived newness. Ironically, in the early fifties I sometimes played chess with Franklin in our faculty club after lunch, where he would generally win as surely as he had beaten me in establishing his result. We never discussed dynamical systems.

Still, the significance that I attached to the result does not show through in the textbook. Nowhere does the expression "unstable in the sense of Lyapunov" appear. As with Poincaré and Birkhoff, everything centers around periodic solutions. The idea of studying irregular solutions of differential equations for their own sake had not really caught on.

Thanks to the interest of Jule Charney, who by then had become a colleague of mine at M.I.T., my paper soon gained considerable attention among meteorologists. Unlike some who saw the paper, Jule believed the results; moreover, he recognized their potential importance for the Global Atmospheric Research Program. One outcome was the predictability experiments performed with the available state-of-the-art global circulation models.

Soon afterward it began to appear that the three equations with their strange attractor, far from being only an abstract device for demonstrating simple chaotic behavior, might yield a fair description of some real-world phenomenon. As a model of ordinary convection, the system is deficient because it places excessive restrictions on the motion. If you set a pan of water on the stove and turn on the burner, the water will not rise everywhere on the left side and sink everywhere on the right, or rise only on the right and sink only on the left, in one big roll, as the equations would require it to do; numerous smaller rolls will develop. If these could somehow be suppressed, the remaining motion might conform more closely to the equations.

One way to suppress the smaller rolls would be to abandon the pan altogether, and let the water fill a tube, arranged in the form of a closed loop, attached to a wall or otherwise held vertical. In a computer simulation, the oceanographer Pierre Welander, then at the Woods Hole Oceanographic Institution, discovered that when such a loop is heated at the low point, at a suitable rate, the water can indeed circulate first in one direction and then in the other, switching at irregular intervals.

The physical reality of the irregularly alternating circulation was substantiated a few years later when the versatile applied mathematicians Willem Malkus and Louis Howard, both at M.I.T. at the time, and Ruby Krishnamurti of Florida State University, temporarily abandoned their pencils and papers, and built water wheels that were specifically intended to execute the behavior predicted by the three equations, spinning clockwise for a while and then counterclockwise for a while. Malkus's wheel was a precision instrument, suitable for controlled laboratory experiments. If you are interested in building your own wheel, you will probably have better luck following Howard or Krishnamurti. The basic ingredients are a turntable—perhaps a plastic lazy Susan—and some paper cups.

Simply make a small hole in the bottom of each cup, near the edge, and then stand the cups up around the periphery of the turntable, with the holes projecting beyond the rim, and glue the cups in place. Figure 47 shows a schematic top view. The smallest circles are the holes, the intermediate ones are the bases and upper rims of the cups, and the large one, shown dotted where it passes under the cups, is the rim of the turntable.

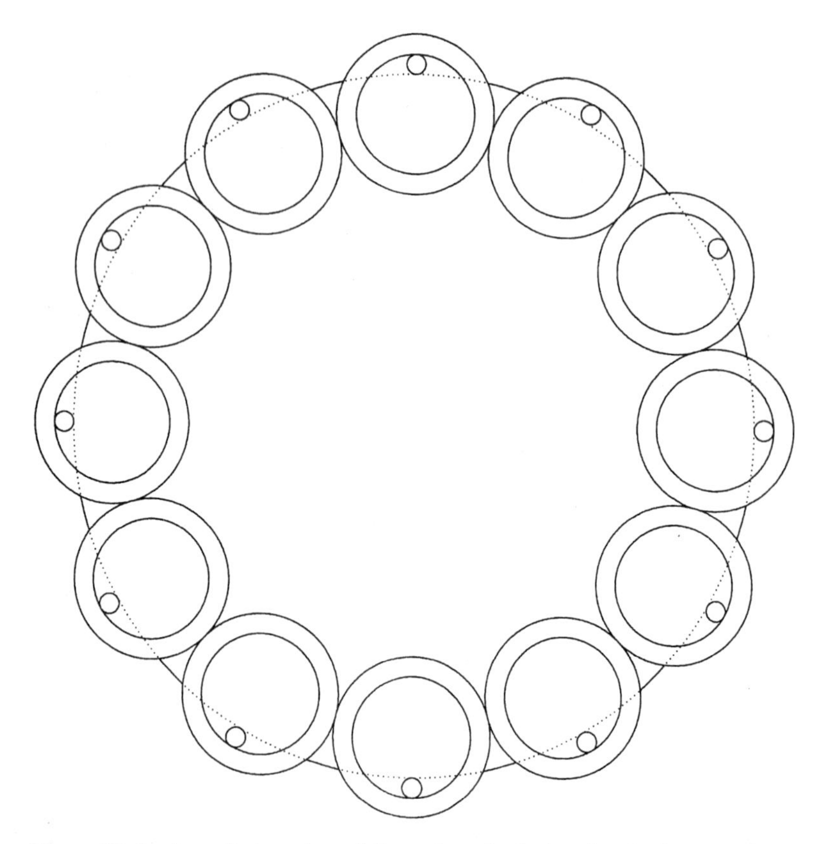

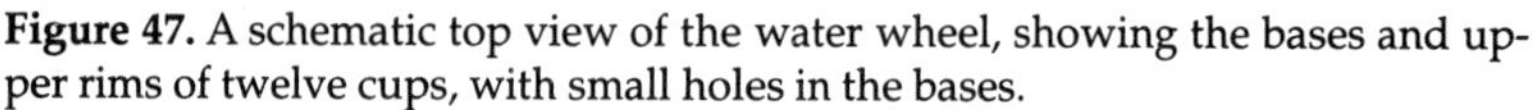

Figure 47. A schematic top view of the water wheel, showing the bases and upper rims of twelve cups, with small holes in the bases.

To operate the wheel, tilt it from the horizontal, perhaps by twenty degrees, and clamp the stationary part of the turntable to something firm. Then let water run at a steady rate from a faucet or a hose into the uppermost cup. As the cup fills, the wheel will become top-heavy and turn, allowing the next cup to move to the top and receive the water, while the first one will slowly lose its supply through the hole. With the wheel tilted at a suitable angle, with the water flowing at a suitable rate, and with a bit of luck, you should see the wheel spin one way, and then the other.

Leaving the laboratory and turning to the library, if you happen to come across a scientific paper that arouses your interest, you may want to look into some related studies, and you can locate some of the earlier ones through the list of references. Unless you are preparing a historical review, you may really be more interested in the later studies, and you may wish that the paper could have looked into the future and listed the papers that would refer to it. Such lists actually appear, after the future is no longer the future, in the *Science Citation Index*, a much-consulted reference journal published by the Institute for Scientific Information, in Philadelphia. It is updated every two months, and the lists are subsequently combined into annual and then five-year cumulative lists.

Scientists in other fields do not routinely read meteorological journals; it is difficult enough to keep up with the literature in one's own specialty. Before 1970, it seems that my paper was cited almost exclusively by other meteorologists. The lone exception, a 1965 paper by the applied mathematician Lee Segel, dealt with thermal convection, a problem not too far removed from meteorology, and Segel's paper would have been accepted for publication by many meteorological journals. From 1970–1974 my paper was cited only two more times by nonmeteorologists.

If I had published the paper in a mathematical journal it might have caught the attention of the chaos community a good deal sooner, but might not have been noticed by many meteorologists. In spite of quips that some scientists take credit for having written two papers when they have really written the same paper twice, that is evidently what I would have had to do to reach a wider audience.

An event that was crucial, for me if not for the development of chaos theory, took place a few years afterward. James Yorke, the mathematician at the University of Maryland whose joint paper "Period Three Implies Chaos" with Tien Yien Li was only the beginning of a long string of innovations, was talking with Alan Faller, who had left Woods Hole and his eight-foot dishpan and had joined the Department of Meteorology at the University of Maryland. Yorke mentioned some work that he was doing, and Faller said that it sounded like my paper on nonperiodicity. He supplied Yorke with a copy, and also made a fair quantity of copies which he sent to various mathematicians and other scientists on the Maryland campus. Shortly afterward, when Yorke was visiting Smale in Berkeley, he showed him his copy, and Smale evidently proceeded to make a number of copies and send them to his acquaintances who were active in dynamical systems—the field was still relatively small. Some, I am sure, saw for the first time a strange attractor in flesh and blood.

In retrospect, it seems to me that what may have distinguished my paper from those that preceded it, other than a picture of a specific strange attractor produced by differential equations, was the idea that chaos was something to be sought rather than avoided. In any event, the "explosion" occurred soon afterward, and my paper was cited on numerous occasions. I would like to think that everyone who referred to it also read it and was influenced by it, but perhaps this is too much to expect. I shall probably never know to what extent it was responsible for setting off the activity that followed, and to what extent I was simply lucky that it became known shortly before an outburst of activity that was due to occur in any case.

The Ubiquity of Chaos

A few years after chaos had burst into prominence, I received a reprint bearing the curious title "Lorenz Knots are Prime." I had not realized that I had any knots, let alone prime ones, which I had never heard of. It was by the topologist Robert F. Williams of Northwestern University, whom I had first met a few years earlier when we were both visiting Steve Smale at the University of California in Berkeley. Topology is a thriving branch of mathematics—one that began as another of Poincaré's creations. Knot theory is in turn a branch of topology.

Topology deals with those properties of curves, surfaces, and more general aggregates of points that are not changed by continuous stretching, squeezing, or bending. To a topologist, a circle and a square are the same, because either one can easily be bent into the shape of the other. In three dimensions, a circle and a closed curve with an overhand knot in it are topologically different, because no amount of bending, squeezing, or stretching will remove the knot. Bob was referring to the butterfly-shaped attractor, and he had shown that some of the closed solution curves that lay in it were knotted.

One of these curves appears in Figure 48 in schematic form; where two branches of the curve cross, the one underneath has been broken. Looking at the figure, I would have guessed that it would unfold into a circle, but, when I duplicated it with a length of cord, joined the ends, and tried to stretch it out, there was the familiar overhand knot, known to knot theorists as the trefoil.

The work was quite specialized, but it typifies the efforts of many topologists who found that they could not resist the lure of strange

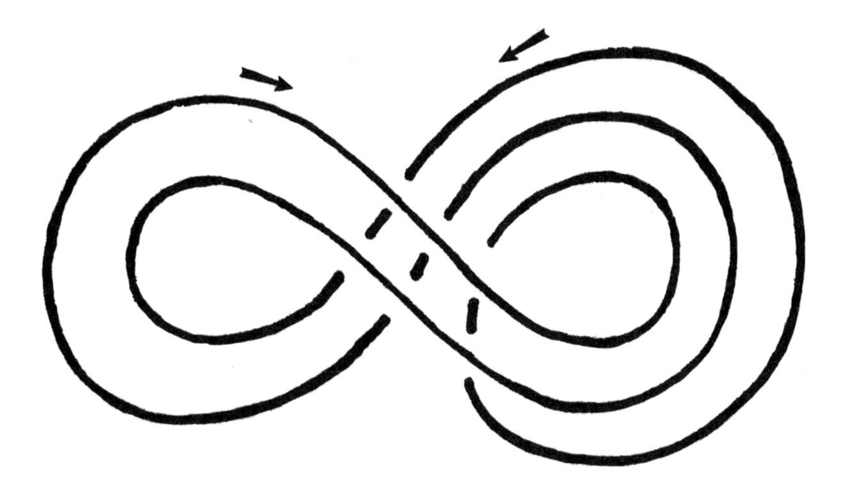

Figure 48. A schematic view of a closed curve contained in the attractor of the three-variable convective model. The curve is in three-dimensional space; where two lines appear to cross, the broken line lies below the solid line. The curve contains a not-so-easily-seen knot.

attractors. They had encountered new aggregates of points, and they would look into such problems as how the innumerable sheets in an attractor are joined together.

Other mathematicians became enamored with bifurcations—the abrupt changes that can take place in the behavior, and often in the complexity, of a system when the value of a constant is altered slightly. We have seen examples in the ski-slope model and the dishpan. Still others turned to chaotic seas with their intricate islands—Hamiltonian counterparts of strange attractors.

Perhaps the more distinguishing feature of the outburst was its spread from one field of endeavor to another. From mathematics, astronomy, and the earth sciences, a general awareness of chaos soon invaded physics, chemistry, and the life sciences, and ultimately the social sciences and the arts. Among the papers that appeared just ahead of the explosion, and served to light the fuse, were a handful that were of special interest to me for one reason or another.

Leading this list is the pioneering work in population dynamics, initiated at Princeton University, by the mathematical biologist Robert May. His dynamical system consisted of a single difference equation in a single variable, closely related to the logistic equation. The population of a particular species, say an insect, often varies greatly from one year to another, and sometimes the total number present during one year is a fairly good predictor for the number during the next. Unlike the pendulum and the sliding board, an insect population has no "Newton's law," and the equations that one formulates generally express what appears reasonable. The conventional assumption is that if the population is very small, it will multiply freely, producing a much larger but still fairly small population the next year. If it is very large, it will produce far more offspring, but food enough to keep them alive will not be available, and again next year's population will be small. The largest population should therefore follow a year with a medium-sized population. May found that for suitable rates of multiplication and starvation, the size would fluctuate chaotically.

A while earlier I had looked into an equation that was somewhat like May's, regarding it as a pure mathematical abstraction. Like the applied mathematicians' water wheels, May's work served as notification that the equation could refer to something tangible.

Also among the early works were a study by Kay Robbins, then a graduate student at M.I.T., dealing with disk dynamos and their role in the reversals of the earth's magnetic field, and one by Hermann Haken at the University of Stuttgart, dealing with lasers. The feature of interest was that in both studies the equations turned out to look very much like mine, and their solutions were much the same. It's not that disk dynamos and lasers are nearly alike, nor that either is much like thermal convection; all three systems had been simplified so much that what was left in one looked much like what was left in either of the others. It is indeed not unusual for simplifications of this sort to lead to the recognition of partial similarities in physically rather different systems, when these likenesses might otherwise go undetected.

One other study left me with mixed feelings. Otto Rössler of the University of Tübingen had formulated a system of three differential equations as a model of a chemical reaction. By this time a number of systems of differential equations with chaotic solutions had been discovered, but I felt that I still had the distinction of having found the simplest. Rössler changed things by coming along with an even simpler one. His record still stands.

Turning to more recent developments, let me again begin by briefly mentioning the life sciences. Outbreaks of various diseases, which to some extent are phenomena in the population dynamics of microorganisms and their victims, have been studied with equations somewhat like those assumed to govern the populations of single species. Often there are two difference equations, one for the predator and one for the host. The heartbeat, with its occasional arrhythmias, has also received considerable attention, which is not surprising in view of its crucial role in our lives. What seems to concern many people as much as the well-being of their hearts, and what might not be expected to be amenable to a similar analysis, is the strength of the local, national, or world economy.

In the past many economists have assumed that the economy has an equilibrium state, and that it would settle down to this state, without any annoying business cycles, if only we would stop meddling with it—in short, if it were not subjected to variable forcing. What some chaos-minded economists are now proposing is that, as a dynamical system, the economy is chaotic, and business cycles, at irregular intervals, are inevitable. Meddling might even suppress rather than produce the cycles, but more likely it would simply shorten some recessions and lengthen others.

Just as the nearly exact equations of meteorology are based on Newton's laws of motion and other laws that refer to the minutest elements of the weather—"parcels" of air small enough to be treated as particles—so any nearly exact equations of economics would have to be based on the far more complicated laws governing the basic economic elements—human beings and some of their creations. Just as synoptic meteorologists have learned from experience how large aggregates of parcels—storms and other structures—typically behave, so economists have learned from experience how various aggregates of people can influence the economy. They have formulated simple systems of equations that incorporate some of the assumed interactions, and in some instances have encountered chaotic solutions.

Turning now to the arts, let us first consider music. Here chaos can enter in two quite different ways. First, there are the tones of musical instruments. A string or a column of air, or to a lesser extent a membrane, usually vibrates with a strong periodic component, corresponding to a fundamental pitch. Typically there are overtones, which contribute to the instrument's characteristic sound, but there is often an irregular component that further modifies the tone, and that in some instances seems to be chaotic rather than truly random. While recently visiting

Douglas Keefe of the Department of Music at the University of Washington, I was rather surprised to learn that the normal tone of the saxophone is not chaotic. Chaos seems to be abundant, however, in a multiphonic tone, produced when the saxophone is played so that two distinct pitches are perceived simultaneously.

Quite a different form of chaos in music is something that has not been detected; it has been introduced during the process of composing. Unless a piece has intentionally been made devoid of structure, there are likely to be some reappearances of earlier themes. These are often more appealing if they are not exact repetitions, but contain a few unanticipated elements. Listen to almost any major work of Brahms—the third movement of his First Symphony, for example—and you will hear him say something and then say it a bit differently the second time, and still differently the third time.

Chaotic solutions of simple systems of equations are noted for their frequent approximate but not exact repetitions. Sometimes more than one "theme" will be repeated; witness Figure 36, produced by an equation of celestial mechanics. Earlier composers had to design their variations, but some of today's composers have, in one way or another, translated the fluctuations of the solutions of simple equations into sequences of notes. Similar steps have been taken in the visual arts.

A short while ago I received an unexpected but welcome package from Carolyn Lockett, then a student in fine arts at the University of Oregon. It contained a video cassette that she had produced. The piece opened with three small bright vertical bars standing against a darker background. The bars soon went into a dance, and for the next four minutes they danced along paths formed by the curves in the butterfly attractor, meeting at times and parting at others. Appropriately enough, she had called the piece *Dance in the Wind*.

Make Your Own Chaos

In the years immediately after my original encounter with the strange attractor, my interest in chaos became centered about its effect upon weather forecasting. When, after ten years or so, I began to receive invitations to speak about the attractor, and learned of some of the things that my hosts were doing, my interest spread to more general aspects. Some of those who were fascinated by the existence of chaos nevertheless felt that it was the exception while regularity was the rule, at least in systems defined by simple sets of equations. Feeling otherwise, I was anxious to disprove the idea. For this purpose I tried to construct other simple systems with nonperiodic behavior. I had little luck in finding anything new. Even as late as 1980, in a paper whose main purpose was to demonstrate the significance of attractors to the meteorological community, I found it easiest to choose a system of equations that could be reduced to the one that I had encountered nearly twenty years earlier.

Soon afterward my luck changed. One system after another that I examined proved to have chaotic solutions. At times it seemed that I could hardly avoid chaos. Equations do not suddenly change their properties, so I must intuitively have changed my method of searching. Chaos was becoming recognized as being ubiquitous, apparently showing up in such diverse phenomena as business cycles and musical tones, but now it appeared to be ubiquitous in another way; systems of equations written down in a rather casual manner had a fair likelihood of behaving chaotically. It is easy now for me to write a prescription according to which you can create your own chaos.

Mappings are easier to design than flows. To avoid having most points go off to infinity, imagine a bounded region, say a square. Determine the result of stretching the square in one direction, say horizontally, compressing it in another, say vertically, and then bending it and fitting it inside the square region that it initially occupied, thereby establishing a mapping of each point of the original square into the point to which it is carried by the stretching, compression, bending, and overlaying. If you are working with more than two variables—presumably on a computer—take a cube or a multidimensional box, and stretch it in at least one direction and compress it in at least one before otherwise deforming it.

As an alternative to bending, you may break the compressed, stretched square into two or more pieces and fit them inside the original square. You will then have a discontinuous mapping. Likewise, you may omit or diminish the compression and fit two or more parts of the bent or broken stretched square over the same part of the original square. You will then have a noninvertible mapping—one in which you cannot always deduce the past state from the present. Of course, you can omit both the compression and the bending, but if you omit the stretching you will not have chaos, since nearby points will not move apart.

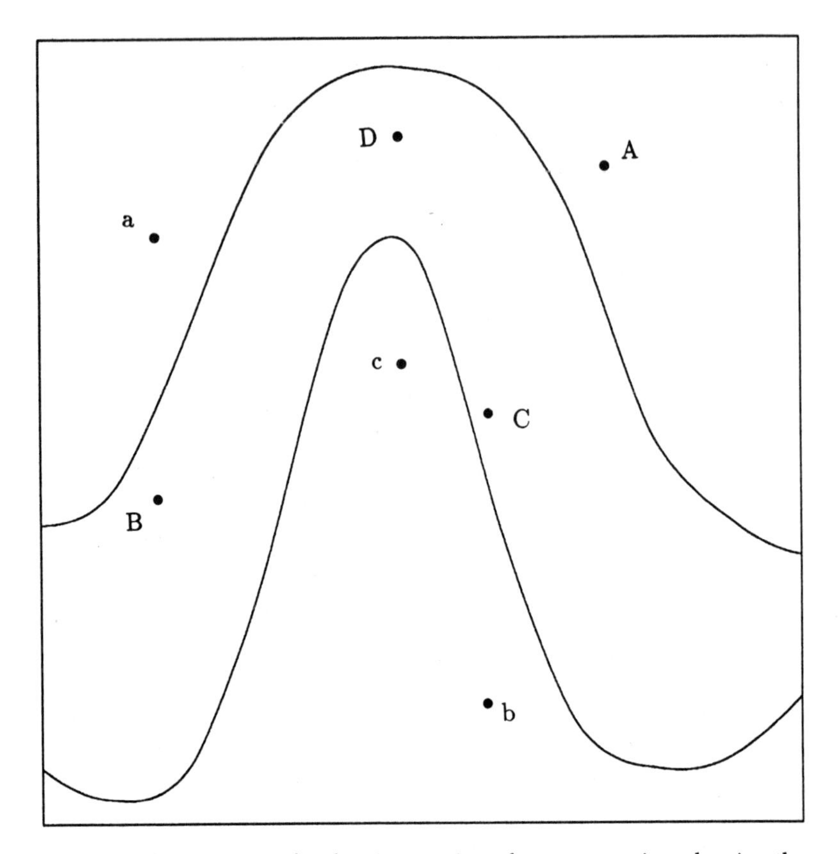

Figure 49. Construction of a chaotic mapping of a square region, showing the square, and the two curves that together with segments of two sides bound the region into which the square is mapped. Points *a*, *b*, and *c* are obtained from *A*, *B*, and *C* by giving the square a quarter turn counterclockwise. Points *B*, *C*, and *D*, obtained from *a*, *b*, and *c* by vertical compression, are the first, second, and third images of *A*.

The mappings that produced the strange attractors in Figures 12 and 16, derived by taking Poincaré sections of the ski-slope models with y as one of the coordinates, are discontinuous, since they are restricted to a fundamental rectangle; when a board or a sled moves out through one side it effectively jumps to the opposite side. A real board sliding on a horizontal plane is a noninvertible system, since, unlike some mathematical boards for which frictional damping is proportional to velocity, it does not move ever and ever more slowly; it comes to an abrupt stop. If we observe the board while it is still moving, we can predict when and

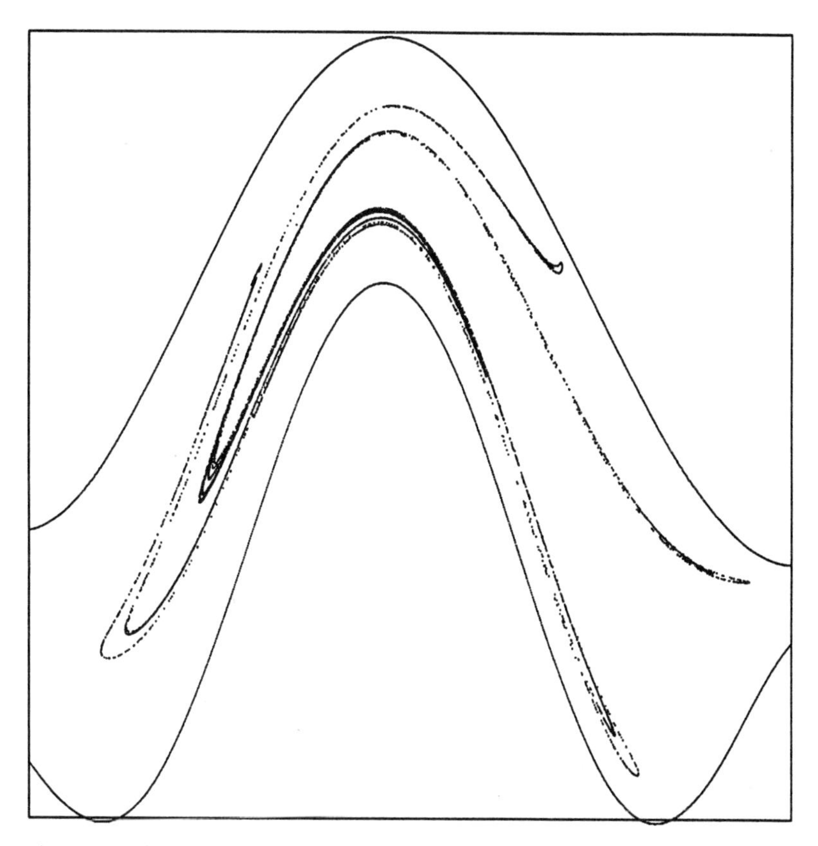

Figure 50. The square of Figure 49, with eighth-degree polynomial approximations to the curves, and the attractor of the mapping, contained between the curves.

where it will stop, and hence where it will be a minute later, but, if we observe it after it has stopped, there is no way to tell when it stopped, and hence where it was a minute ago.

To construct a special continuous invertible mapping in two dimensions, take a square and draw two curves extending from one side to the other, as illustrated in Figure 49. The curves should not cross each other, nor should they cross the top or bottom of the square, and neither curve should intersect any vertical line more than once.

To produce the sequence of points into which a given point, say point A in the figure, is successively mapped, first give the square a quarter turn counterclockwise. This will carry point A to point a. Alternatively,

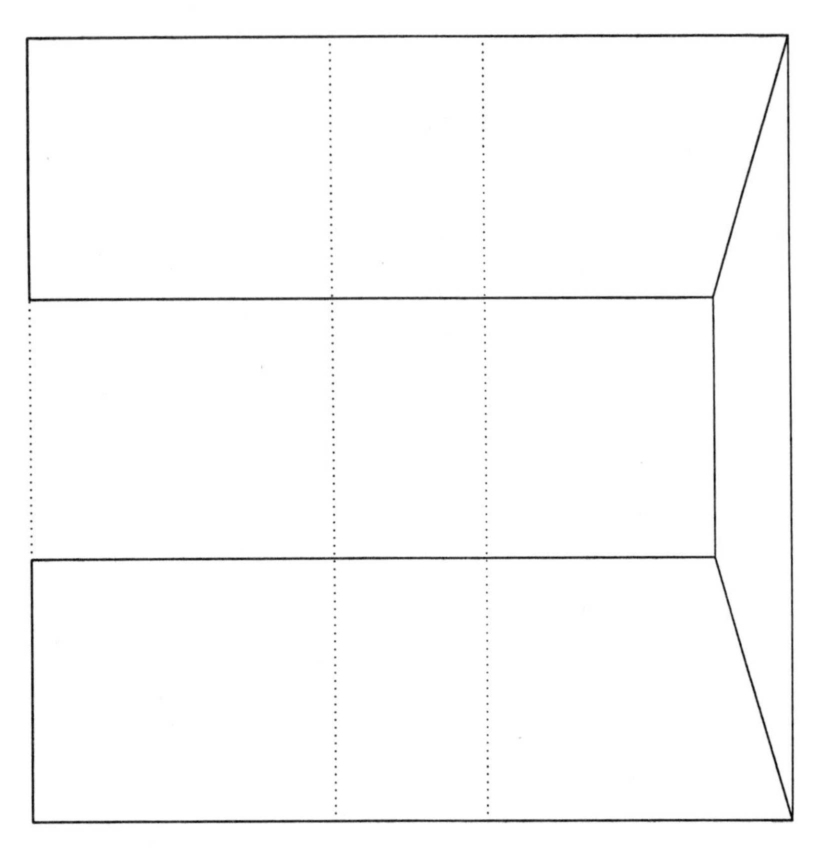

Figure 51. Construction of a chaotic mapping of a square region, showing three sections of the square separated by dotted vertical lines and the three trapezoids into which these sections are mapped.

and with a different result, you could have turned it clockwise, or flipped it over one diagonal or the other. Second, compress each vertical line in the square so that it fits between the two curves. For example, if a is three-fourths of the way from the bottom to the top of the square, replace it by point B, on the same vertical line and three-fourths of the way from the lower to the upper curve. Point B is the point to which A is mapped. Repeating the two steps will produce points b and C, c and D, etc. Note that this is somewhat like the process that produced the horseshoe in Figure 41.

The procedure is not guaranteed to produce chaos. You are more likely to succeed if your curves have some rather steep slopes. Note that

Figure 52. The attractor of the mapping of Figure 51.

both the stretching and the shrinking take place in the second step. Points originally separated horizontally, and hence separated vertically after the rotation or flipping, will move closer together, but points originally separated vertically will move farther apart. Chaos can result if moving apart prevails over moving together.

Unless you work with a high-quality drafting set, you are likely to misplace a point, after which the succeeding points will be even more badly misplaced, and, in any case, you may find the procedure rather tedious after locating twenty points or so. These might be enough to suggest sensitive dependence, if you compared them with a second sequence of twenty points, similarly found. However, they will probably be insufficient to establish any lack of periodicity, and they will be totally inadequate for revealing the strangeness of the attractor, for which

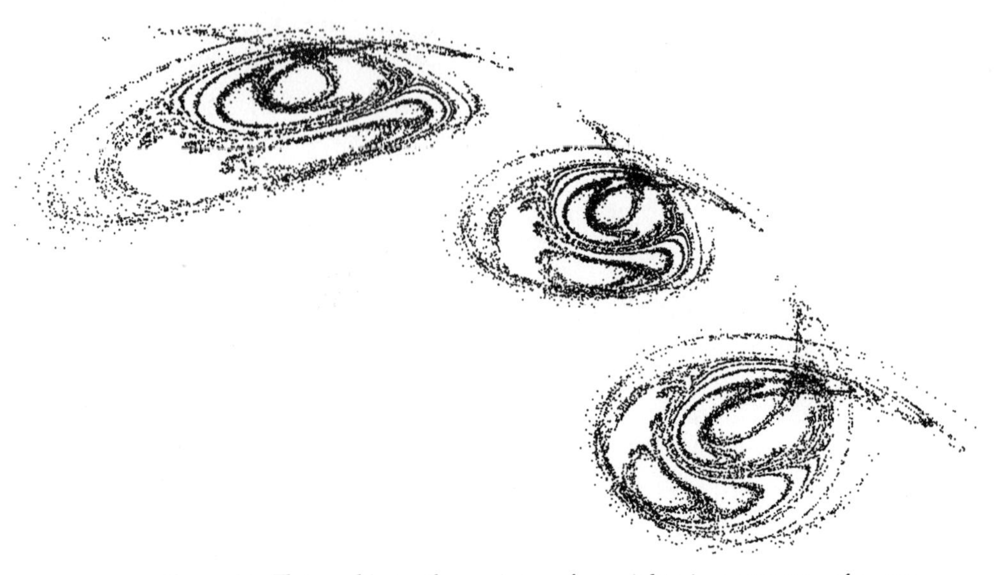

Figure 53. Flying objects: three pieces of an eight-piece attractor of a two-dimensional noninvertible mapping.

hundreds and probably thousands of points will be needed. Abandoning the drafting set and turning to a computer is therefore strongly recommended. For the curves, you will need formulas that express height above the base in terms of distance from one side.

The curves in Figure 49 were drawn by hand. In Figure 50 the same curves have been approximated, apparently not too closely in some spots, by formulas that the computer can easily handle. Point A has again been chosen as a starting point, and a sequence of 10,100 points has been generated. The first 100 points have been discarded as possibly representing transient conditions, and the remaining ones, plotted in the figure, clearly reveal a strange attractor, which of necessity lies between the two curves.

If you prefer straight lines, you can turn to a variant of Smale's horseshoe, which avoids the question of what happens to points whose images leave the square, by keeping all the images inside. Let the dotted vertical lines in Figure 51 divide the square into three strips, while the solid lines divide a portion of it into three trapezoids. Compress the left-hand strip vertically, and then stretch it horizontally, more at the bottom than at the top, so that it just fits into the lower trapezoid. After compressing the middle strip similarly, give it a quarter turn counterclockwise before stretching it vertically to fit into the right-hand trapezoid. Give the compressed right-hand strip a half turn before making it fit into the upper trapezoid. The widths of the strips and the trapezoids can be anything desired. On the computer you will need three pairs of equations—one for each strip.

Again, chaos is not guaranteed. I was unsuccessful on my first few tries. The trick seems to be to make the right-hand trapezoid very narrow—perhaps one-tenth as wide as the square. Figure 52 shows the strange attractor produced by a successful try, and even here there are some surprises; short dark streaks with virtually all possible orientations pop up in one place or another.

I cannot give you a prescription for using the computer to make the most interesting attractors, but by now you should be able to proceed on your own. You will undoubtedly find mappings easier to work with than flows. You can produce some new effects by choosing noninvertible mappings. Here is one of my favorite examples, in Figure 53. The three flying objects, though strange, are by no means unidentified; for identification, see the leading section of Appendix 2.

Is Randomness Chaos?

The collection of phenomena that we recognize as behaving chaotically has become so great that it would be hard to compile a comprehensive list. I have only sampled it. There are basically two sets of circumstances under which it can become still larger. We may believe that some phenomenon is governed by deterministic laws and that it responds in a regular manner, only to discover at some point that its behavior is more irregular than we suspected. The motions of some of the heavenly bodies, considered to be quite regular before Poincaré's discoveries, constitute a classical example. We may, on the other hand, be quite aware of the irregular behavior of a phenomenon, but we may think that this behavior results from some inherent randomness, and discover only later that the phenomenon obeys regular laws. Some people would place breaking ocean waves in this category. Here I am adopting the more liberal concept of chaos, and am including processes with some true randomness, provided that the processes would exhibit similar behavior if the randomness could be removed.

The question that I wish to address now is whether chaos is so ubiquitous that all or most of the processes that we still regard as behaving randomly should become recognized as being chaotic instead. Basically, this is the question that Poincaré asked in his essay on chance. Perhaps there is little to add to what he said, but the question may be worth revisiting in the light of our knowledge that chaotic phenomena are so abundant.

Before proceeding further, we need to consider the question of the free will of human beings, and perhaps of other animate creatures. Most of us presumably believe that the manner in which we will respond to a given set of circumstances has not been predetermined, and that we are free to make a choice. For the sake of argument, let us assume that such an opinion is correct. Our behavior is then a form of randomness in the broader sense; more than one thing is possible next.

Are all irregularities, except those produced by intelligent behavior, chaotic rather than random? A simple example suggests the possibility. Suppose that we have paused to admire a maple tree whose leaves have assumed a brilliant golden hue. Every once in a while we may see a leaf floating irregularly from the branches of the tree down to the ground. Are we looking at randomness or chaos?

Very likely a gust of wind will first detach the leaf from the tree. The branches may also be swaying, so that the wind will determine the position of the branch when the leaf begins to fall. Even if the leaf becomes detached for some other reason, the wind will guide it in its journey to the ground, presumably according to the laws of aerodynamics.

The wind that blows past the tree during the quarter of a minute or so that the leaf is falling is a part, albeit a very small part, of the global weather system. If we admit that the weather is an instance of chaos, we are more or less forced to say the same thing about the wind and then the leaf. A while before the leaf fell, perhaps only a minute but perhaps a day or more, some human activity, which is certainly more powerful than butterfly activity, presumably altered the course of the ensuing weather, including the wind that carried the leaf. We have agreed, however, that as long as the weather would have behaved equally irregularly without our interference, we are looking at chaos. Note that our past activity, by changing the wind, will change the path of the leaf, but it will change it for another path that might equally well have been anticipated. I suspect that many other seemingly random phenomena that do not depend primarily on animate activity for their apparent randomness can be analyzed in a similar way.

Of course I have left out a gigantic class of random phenomena—the ones that occur on subatomic scales and are governed by the laws of quantum mechanics. Here it is a fundamental premise that events occur at randomly spaced discrete times. Since all matter is ultimately divisible into subatomic particles, does this mean that all matter behaves randomly, and determinism is only an abstraction?

Perhaps so, but chaos should remain if we again adopt the liberal interpretation. I suspect that the general behavior of the swinging pendulum, the rolling rock, the breaking wave, and most other macroscopic phenomena would not be noticeably altered if quantum events occurred at regular predictable instants, or at chaotically determined instants, instead of randomly.

Let us turn to the alternative possibility—that the future course of the universe, including the animate activity within it, has already been determined, and that our apparent free will is an illusion. It may surprise us to learn that anyone could take such a suggestion seriously, but the idea has been proposed time and again over the centuries by various philosophers. It is probably most often associated with the French mathematician Pierre Simon de Laplace, who lived a century before Poincaré.

I have encountered the suggestion that, if things are completely deterministic, we should alter our view of our fellow beings, and should not, for example, punish a murderer or some other criminal, because it was already determined that he was going to commit the act and there was nothing that he could do about it. Such a proposal fails to capture all of the implications of determinism. If it has already been determined that someone will commit a murder, then, by the same token, it has already been determined whether or not we shall punish him, and there is nothing that we can do about it. Indeed, it has even been predetermined that some of us will not believe in predeterminism and will think that we can do something about it.

What, then, should we choose to believe—that everything has been determined, or that we are free to make decisions? I believe that the appropriate answer is obvious if, like mathematicians, we introduce certain premises before attempting to reach conclusions. Let our premise be that we should believe what is true even if it hurts, rather than what is false, even if it makes us happy.

We must then wholeheartedly believe in free will. If free will is a reality, we shall have made the correct choice. If it is not, we shall still not have made an incorrect choice, because we shall not have made any choice at all, not having a free will to do so.

CHAPTER 5

What Else Is Chaos?

Nonlinearity

AT ABOUT THE TIME that investigations of chaos were becoming fairly common, the well-established journal *Physica,* which had already acquired some of the aspects of three separate journals, inaugurated a fourth series, *Physica D,* to be devoted to nonlinear phenomena. This move took place a few years after the founding of another journal, *Nonlinear Analysis: Theory, Methods and Applications.* Several years later volume 1, number 1 of *Nonlinearity* appeared. Contributors to these journals represented many disciplines, and came from many institutions, including the Center for Nonlinear Studies at the Los Alamos National Laboratories, the Center for Nonlinear Dynamics at the University of Texas, and the Institute for Nonlinear Science at the University of California. What did their contributions deal with? In many instances it was chaos.

Are "nonlinearity" and "chaos" synonyms? Not at all. First of all, the words differ in that the former has a single meaning. A *linear* process is one in which, if a change in any variable at some initial time produces a change in the same or some other variable at some later time, twice as large a change at the same initial time will produce twice as large a change at the same later time. You can substitute "half" or "five times" or "a hundred times" for "twice," and the description will remain valid. It follows that if the later values of any variable are plotted against the associated initial values of any variable on graph paper, the points will lie on a straight line—hence the name. A *nonlinear* process is simply one that is not completely linear.

Perhaps the most familiar truly linear processes are those that we ourselves have created. A simple example is the purchase of food or other items, provided that no discount is offered for buying in large quantities. If we buy a dozen eggs and leave the store with one dollar less than we would have if we had not bought the eggs, someone else buying two dozen eggs of the same size and quality will leave with two dollars less.

Just as few concrete physical systems are strictly deterministic in their behavior, so very few are strictly linear. The great importance of linearity lies in a combination of two circumstances. First, many tangible phenomena behave approximately linearly over restricted periods of time or restricted ranges of the variables, so that useful linear mathematical models can simulate their behavior. A pendulum swinging through a small angle is a nearly linear system. Second, linear equations can be handled by a wide variety of techniques that do not work with nonlinear equations.

It is easy to see, without resorting to extensive mathematical analysis, that the ski-slope model is nonlinear. Consider a board that starts from rest at a point 1 meter due east of a mogul. It will begin to slide in the direction of steepest descent, and so will move more or less southeastward. Consider a second board 5 meters east of the same mogul. It will behave just as the first, since it is 1 meter east of the next mogul.

If the system were linear, a third board starting midway between the first two would have to remain midway between them and would also move southeastward. However, since it would be starting 1 meter *west* of the second mogul, it would actually move more or less southwestward.

The same sort of reasoning shows that any chaotic system must be nonlinear, provided that each variable of the system is confined within certain limits. Such a variable might be the temperature at Phoenix, which, regardless of the way some residents may feel, cannot approach the boiling point of water, nor can it drop to the depths often encountered in central Alaska. If the global weather pattern is really sensitive to the flap of a butterfly's wings, so that a time will finally arrive when the temperature at Phoenix will be 10 degrees higher than it would have been without the butterfly's aid, a disturbance 100 times as great, which could easily be produced by the flap of a sea gull's wings, would, if the weather behaved linearly, have to produce a temperature at Phoenix 1000 degrees higher—a clear impossibility. Even if the weather is not sensitive to butterflies, the argument is still valid; a chaotic system is one in which small differences in the present state will lead in due time to the largest differences that can occur. If such a system were not nonlinear, moderate-sized differences of a suitable form would have to be followed by differences too large to occur.

Even though chaos demands nonlinearity, nonlinearity does not ensure chaos. A single example constitutes sufficient evidence, and one example is the now familiar ski slope, with the height of the moguls above the adjacent pits reduced to 50 centimeters. Here we have seen that a board, after perhaps weaving back and forth for a short time, will continue to progress southeastward forever, or else southwestward forever, in a regular periodic manner. A slight disturbance in the state will produce a new path, which, however, will soon converge upon the original one.

It may not seem particularly surprising that the qualitative behavior of a system can change when the intensity of some disturbing influence passes a critical level. In looking at bifurcations in more detail, however, we saw that chaos on the ski slope could return again with 43-centimeter moguls, and would disappear again at 40 centimeters. Behavior of this sort is but one example of the surprises that may be waiting for us when we enter the realm of nonlinearity.

Complexity

On the following pages we see three figures, two produced by the computer and one a photograph. Which one strikes you as being the most complex?

If you and a friend compare answers and do not agree, this is no cause for concern. The term "complexity" has almost as many definitions as "chaos."

It would seem hard for anything to be much more complex, in the nontechnical sense of being made up of numerous parts, than the system of which Figure 54 shows but a tiny portion. This fascinating system is the *Mandelbrot set;* in the figure it is accompanied by enough surrounding points to render it visible. Augmented by still more surrounding points, it is familiar to many in its spectacular multicolored form, where it shows up as an intricately woven assemblage of ripples, stars, jewels, and sea horses.

The Mandelbrot set is a fairly simple mathematical concept. To draw a picture of it, we let one location on a plane be the "origin," and choose

Figure 54. A small portion of the Mandelbrot set, with enough surrounding points to render it visible. The region extends horizontally from -0.664 to -0.634 and vertically from -0.498 to -0.468.

another location for a "key point." We then let a point start at the origin, and "jump" about on the plane, in accordance with a special rule that specifies the spot to which the point will jump in terms of the key point and the spot from which the point has jumped; the rule appears in the discussion of the logistic equation in Appendix 2. The first jump, incidentally, takes the point to the key point. If the point remains within a restricted region surrounding the origin, the key point lies in the Mandelbrot set, and we plot it. We then repeat the procedure for each of a large collection of key points, without changing the origin or the rule.

One might imagine that a point whose key point is just outside the set would jump in a manner much like one with a nearby key point just in-

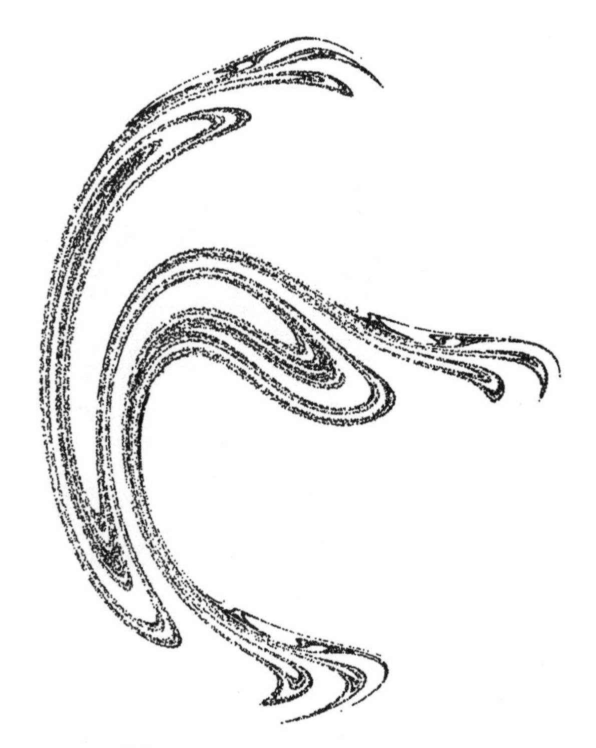

Figure 55. A variant of the Japanese attractor.

side, for a while at least. This is indeed the case; points whose key points are just outside require many jumps to exit from the restricted region. To construct the figure I took two million key points, chosen randomly from the region covered by the figure, and I plotted those key points, 83,423 in all, whose associated jumping points had not exited after sixty-eight jumps.

What might be even more complex than the Mandelbrot set? Perhaps the entire global weather pattern—perhaps the anatomy of a single human being.

Just as there are centers dedicated to studying nonlinearity, so there are the Center for Complex Systems Analysis at the University of Illinois and the Complex Systems Theory Branch of the Naval Research Laboratory in Washington. Just as there are journals devoted to nonlinearity, so there has been since 1978 the journal *Complex Systems*. Just as studies in nonlinearity often deal with chaos, so studies in complexity

Figure 56. Some wind streaks in a field of packed snow.

often deal with chaos. Indeed, *complexity* is sometimes used to indicate sensitive dependence and everything that goes with it.

By this definition Figure 55 is the most complex. It is one of a family of strange attractors discovered by Yoshisuke Ueda of Kyoto University. Like the Cartwright-Littlewood attractor of Figure 32, it is a Poincaré section of the attractor of a periodically forced dissipative system—this time the so-called Duffing oscillator—but here the visible resemblance ends. Ueda had found evidence of strange behavior as early as 1961, but a high-resolution picture had to await a more powerful computer. He has recently presented a vivid description of the difficulties that he encountered in getting his ideas accepted. In a summary article, David Ruelle described Ueda's original attractor as the most aesthetically pleasing strange attractor so far produced—Figure 55 is somewhat grotesque by comparison—and he referred to it as a Japanese attractor, but, on a recent visit to Ueda's laboratory in Kyoto I learned that it was being called *the* Japanese attractor.

By contrast, the Mandelbrot set is not a consequence of complexity, in the sense of chaotic behavior. Most of the jumping points whose key points are in the set show no sensitive dependence in their jumping, and they soon jump about in periodic sequences. The only exceptions are the points whose key points are on the boundary of the set.

Sometimes a distinction is made between "chaos" and "complexity," with the former term referring to irregularity in time, and the latter implying irregularity in space. The two types of irregularity are often found together, as, for example, in turbulent fluids.

Complexity is frequently used in a rather different sense, to indicate the length of a set of instructions that one would have to follow to depict or construct a system. By this measure, Figure 56 is the most complex. It looks somewhat like part of a strange attractor—the dark sinuous curves are separated by wider light spaces, and some apparent curves may be seen on closer inspection to be pairs of curves. However, we are actually looking at wind streaks in the packed snow that blanketed the well-frozen Sudbury River in eastern Massachusetts one morning in 1977.

Figures 54 and 55 can be readily reproduced by following short sets of instructions or executing short computer programs, but the streaks do not conform to any known simple mathematical formula. To reproduce them numerically we would have to specify the locations of so many points on each streak that everything else could be filled in by interpolation. If we were looking at striations in a ledge, we might think of letting the instructions specify the precise latitude and longitude to which a camera should be taken, but streaks in the snow soon disappear, and these particular ones were nowhere to be seen when my wife and I made our next ski journey along the river a few days later.

Fractality

There are some quantities that can be measured only in whole numbers—the number of children in a family, even though a mother expecting her third may say that she has two and a half, the number of runs scored in a baseball game, the number of letters in a word. Most people would also include the number of dimensions of an object or figure; a three-dimensional ball can have a two-dimensional shadow with a one-dimensional outline. Near the end of the nineteenth century, however, mathematicians appeared to discover that some sets of points had fractional dimensionality. What really happened, of course, was that some pioneers investigated certain structures that the majority had regarded as weird, and they found that these weird structures lacked some of the properties generally associated with simple geometrical objects of one, two, or more dimensions. They then redefined dimensionality in such a way that curves, surfaces, and solids remained one-, two-, and three-dimensional, while precise values of dimension could be assigned to the less familiar structures. In many cases these values proved to be fractions. Subsequently other seemingly logical definitions were introduced; the different definitions have not always agreed.

A frequently quoted definition is the one introduced in 1919 by the German mathematician Felix Hausdorff. A modified definition, which I find easier to comprehend, and which is equal to the Hausdorff dimension for many sets of points but larger for others, is the *capacity*, introduced by the Russian mathematician Andrei Kolmogorov. I shall be referring to it as the dimension. It is most easily illustrated when the points of the set lie on a plane, and are bounded by a square, say one meter on a side.

We divide the square into four squares, each half a meter on a side. We then divide each of the new squares into four squares, one-quarter of a meter on a side, etc. At each step we note the number of newly obtained squares needed to cover all the points of the set. We are interested in how this number increases from one step to the next after many steps.

If the set of points whose dimension is being evaluated is the whole interior of the square, or one or more filled-in areas, the number of squares immediately or eventually quadruples at each step; it increases by a factor of 2^2. If it is a curve or a finite number of curves of finite length, the number in due time simply doubles, because most of the new small squares will lie between curves; it increases by 2^1. Finally, if it is a single point or a finite set of points, the number eventually ceases growing altogether; it increases by 2^0. In each case, the exponent of the factor, 2, by which the diameter of a square is reduced at each step, is the dimension of the set.

We now ask whether there are sets for which the number of squares ultimately goes up as 2^d, where d is not a whole number. If so, the dimension of the set, defined in any case to be d, is fractional.

Such sets are not hard to create. Consider, for example, the result of starting with a single square, and then successively dividing every square into four smaller squares and discarding each upper-right new square as it is produced. The first step will produce a fat letter "L," the second will produce a staircase with four steps and a square hole, and the end result, seen in Figure 57, will be a set of line segments forming nested isosceles right triangles. To cover these segments with squares we would have to triple the number of squares at each step. It follows that if d is the dimension, $2^d = 3$, so that $d = 1.586$.

If mathematicians nearly a century ago dealt with fractional dimensions, it remained for Benoit Mandelbrot to recognize in the middle twentieth century that sets with fractional dimensions need not be weird, and that in fact many of the familiar systems found in nature or in everyday life, or at least simple mathematical models of these systems, have fractional dimensions. Such systems include trees with their trunks and limbs and branches and twigs, and mountains where smaller and smaller features can be rougher and rougher. Indeed, one of his best-known earlier papers is entitled "How Long Is the Coast of Britain?" What he noted was that, if one measures the length on successively larger maps that resolve successively smaller features, the answer becomes larger and larger. This is equivalent to covering the coastline with smaller and smaller squares, and deducing that its dimension is somewhere between 1.0 and 2.0; Mandelbrot has suggested 1.25.

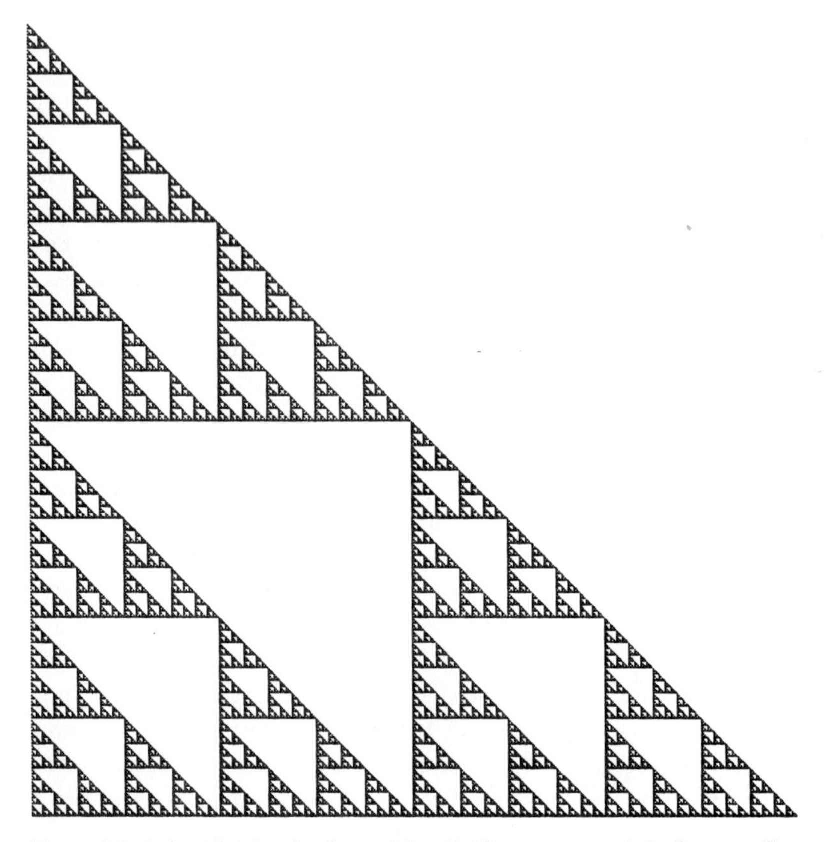

Figure 57. A fractal triangle, formed by dividing a square into four smaller squares and discarding the upper right square so produced, then dividing each remaining square into four still smaller squares and discarding each upper right square so produced, and repeating the process indefinitely.

Mandelbrot coined the term *fractal* to describe systems with fractional dimensionality. The coastline of Britain and the triangular structure in Figure 57 are fractals. Unlike many coined words, the term immediately became widely used, and, unlike "chaos" and "complexity," it still appears to have essentially one meaning.

A property of many fractals that Mandelbrot has repeatedly discussed in his papers is *self-similarity:* in many fractal systems, several suitably chosen pieces, when suitably magnified, will each become identical to the whole system. This implies, of course, that several subpieces of each piece, when magnified, become equivalent to that piece, and hence to

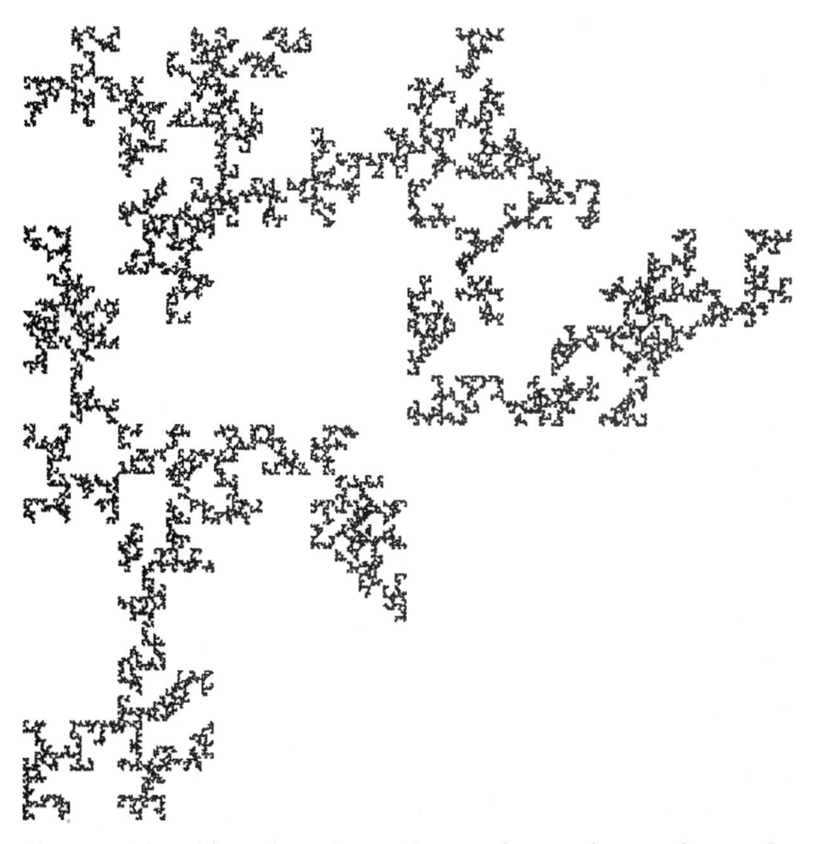

Figure 58. A fractal formed as in Figure 57, except that in each retained square the corner to be discarded has been chosen randomly.

the whole system. In Figure 57 each small triangle is clearly identical in structure to the large one. Other fractals are only statistically self-similar; small pieces, when magnified, will not superpose on the entire system, but they will have the same general type of appearance. Such a fractal appears in Figure 58, constructed in the manner of Figure 57, except that, instead of the upper right corner, a randomly chosen corner of each square has been removed.

By using the concept of self-similarity you can construct your own fractal tree. If you have a drafting set or if you are satisfied with a rough sketch, you don't even need a computer. Begin with a vertical line segment; this is the trunk. From the top of the trunk, draw line segments extending horizontally to either side, each six-tenths as long as the trunk;

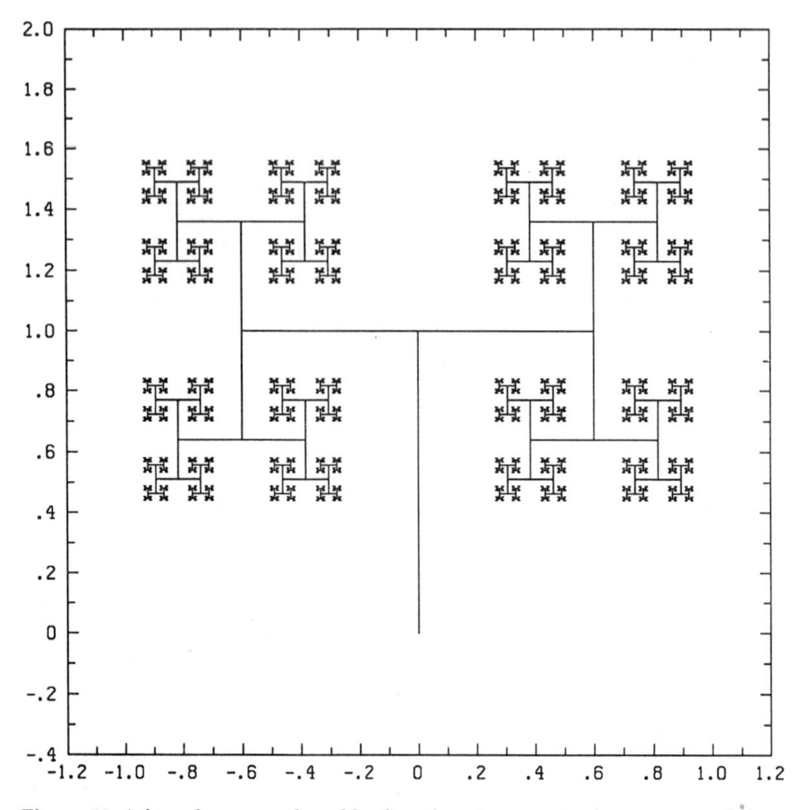

Figure 59. A fractal tree, produced by first drawing a vertical segment, and then, after this segment or any other one has been drawn, treating it as a "parent" segment and drawing two "offspring" segments, each six-tenths as long as the parent, and each extending at right angles from the end of the parent.

these are the limbs. From the end of each limb, draw a segment extending upward and one extending downward, each six-tenths as long as a limb. Continue, alternating between horizontal and vertical segments, until the lines are as close together as the width of the pencil marks. You will obtain something like Figure 59. If you could carry out the process for an infinite number of steps, you would have a fractal, with dimension 1.356. The leaves, that is, the points to which the branches eventually converge, also form a fractal with dimension 1.356.

The ratio of limb length to trunk length need not be six-tenths. If you reduce it below one-half, the leaves will form a fractal with a dimension smaller than 1.0, even though the tree will look much like the one in Figure 59. If you increase the ratio to 0.707, i.e., $\sqrt{1/2}$, the dimension will reach 2.0 and the tree and also the leaves alone will fill in a rectangle. There is no way that a drawing on a plane can have a dimension greater than 2.0, and if you make the ratio greater than 0.707, the shorter branches will simply fall on top of the longer ones infinitely many times.

What if you make the ratio exactly one-half? The dimension will be exactly 1.0, but the tree will still look much like Figure 59. It seems reasonable to include sets like this in the family of fractals, even though their dimensions "happen to be" whole numbers.

You may object that the figure looks more like an overloaded telephone pole than a tree. There are other options. At each branching the two smaller branches can leave the parent branch at different angles, not necessarily right angles, and their lengths can bear different ratios to the length of the parent branch. It would also be possible to have three or more limbs leaving the trunk at different heights. The more variations you include, the better chance you will have of getting something that looks like a tree.

Figures 60 and 61 show two attempts. They were produced by the same computer program that produced Figure 59. The only differences were the choices of the two angles and the two ratios; you can see what these are by comparing the leading limbs with the trunk. Interesting structures, not always leaflike, can also be created by plotting only the leaves, as in figures 62 and 63. The pictures would seem to support Mandelbrot's claim that fractals, or at least fractals as seen with finite resolution, are abundant in nature.

What do fractals have to do with chaos? Very little, when the fractal is the triangle produced by removing upper right corners from squares. There is nothing in Figure 57 that looks like randomness. The most that can be said is that a simple routine has produced a figure that would look rather strange to anyone who had viewed only the figures that appear in mid-twentieth-century geometry books.

The triangle might even be called chaos in reverse, in view of an alternative procedure for constructing it, which I first heard about in a lecture presented by Michael Barnsley of the Georgia Institute of Technology. Locate the three corners and plot a point anywhere on one side. Choose one corner at random, and plot the point midway between the first point and this corner. Again choose a corner at random, and plot the point midway between the new point and this corner. After enough repeti-

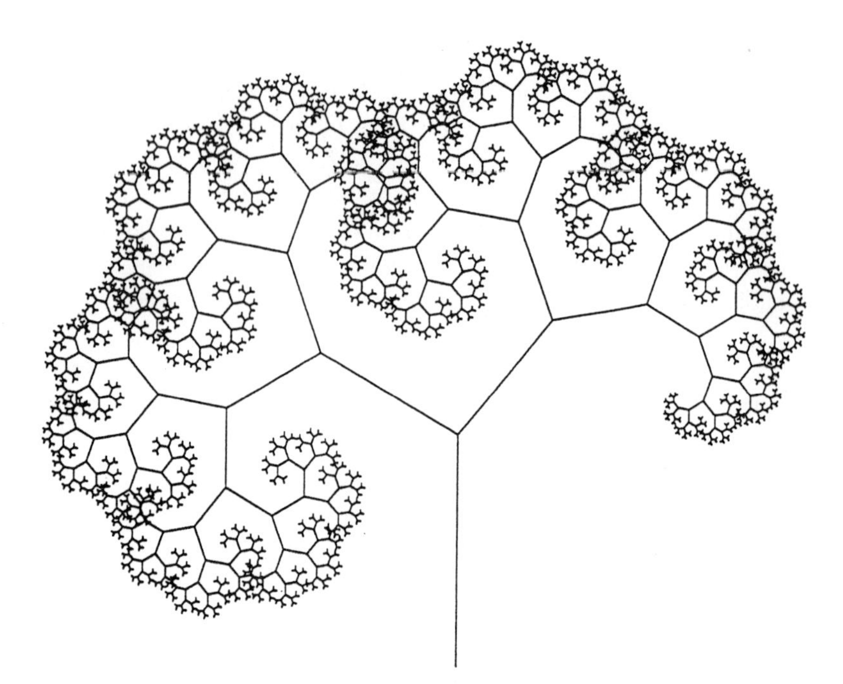

Figure 60. A fractal tree drawn as in Figure 59, except that the ratios, instead of both being 0.6, are 0.7 and 0.65, and the angles, instead of being -90° and +90°, are -60° and +40°.

tions you will reproduce the triangle in Figure 57. The remarkable thing is that if you choose the corners in a regular sequence, instead of randomly, you will end up with just a few points. Thus, in a real sense, the triangle is something that is random but does not look random.

While I am talking about the triangle, let me mention another way to produce it. Take a large piece of paper marked off into small squares. Place an "x" or some other mark in one square in the top row. In the next row and then in subsequent rows, place a mark in a square if the square just above it or the square just to the left of the one just above it has a mark, but not if both have marks. After continuing for a great many rows, and then looking at the paper from a distance so that the marks appear to merge, you will see the triangle.

The process by which each row will have evolved from the one above it is actually a dynamical system—a mapping—rather unlike any others that we have seen. It has an infinite number of variables, represented by the squares that form a row, but there are only a finite number of values

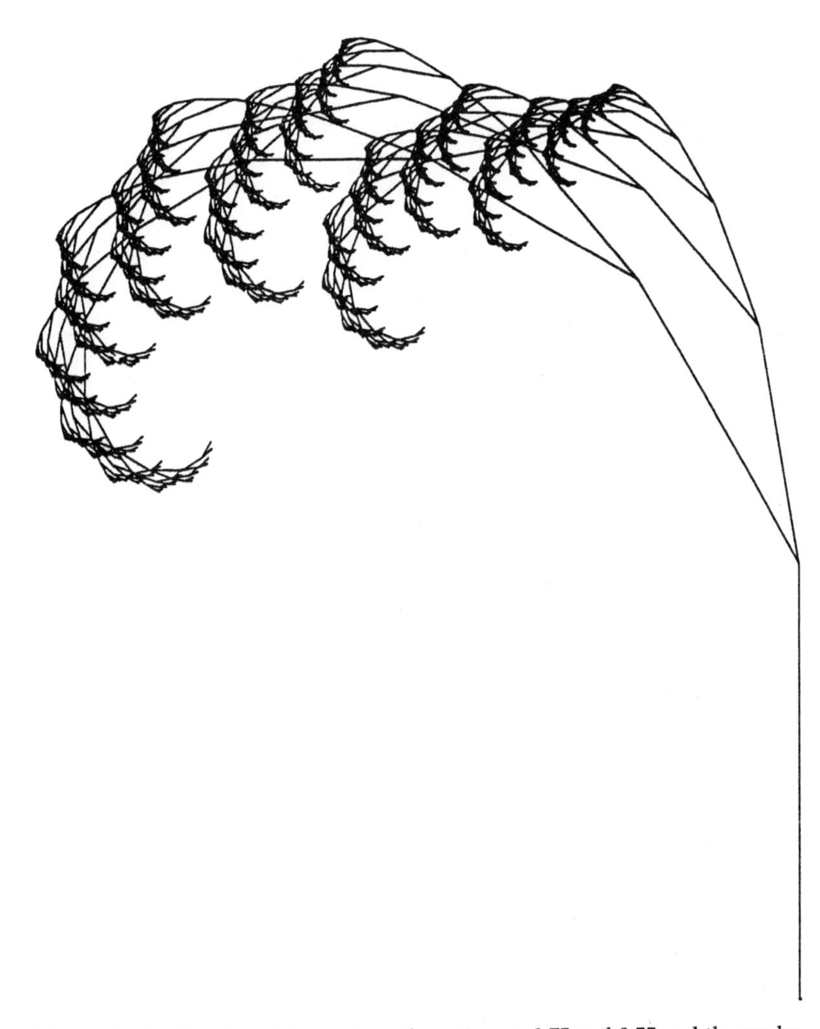

Figure 61. Another fractal tree, where the ratios are 0.75 and 0.55 and the angles are -30° and -10°.

that any variable can assume—only two in this case, symbolized by an "x" and a blank square. Progression from row to row represents advancement in time. Such systems are called *cellular automata,* and they first entered mathematics as another product of John von Neumann's fertile imagination. Interest in them has flourished partly because multidimensional automata appear to afford an especially economical means of simulating fluid motion.

Figure 62. The leaves of a fractal tree, where the ratios are 0.7 and 0.65 and the angles are -80° and +20°.

Some fractals come close to qualifying as chaos by being produced by uncomplicated rules while appearing highly intricate and not just unfamiliar in structure. There is, however, one very close liaison between fractality and chaos; strange attractors are fractals.

For attractors like those in figures 12 and 16, with their complexes of nearly parallel curves, the number of small squares needed to cover every point will more than double as the side of a square is cut in half, since each square will extend across fewer curves. It will not quadruple, however, since a larger and larger fraction of the squares will fall between the curves. The dimension therefore lies between 1.0 and 2.0.

Figure 64 has been constructed so that every vertical line drawn through it intersects it in a Cantor set whose dimension, between 0.0 and 1.0, is indicated at the base of the figure. It may therefore be used as a scale for estimating the fractional part of the dimension of certain struc-

Figure 63. Another assemblage of leaves, where the ratios are both 0.71 and the angles are -70° and +70°.

tures like those in figures 12 and 16, whose intersections with straight lines are Cantor sets. One can observe whether these sets are rather sparse, as on the left side of the scale, or rather dense, as on the right. For dimensions of 0.1 or even a bit higher, only two intersections with a vertical line will be resolved by a drawing; this would also be true of a line intersecting the butterfly attractor, which looks as if it consisted of just two merging surfaces. Lines cutting across Figures 12 and 16 would look more like the central portion of the scale, and the attractors can be estimated to have dimensions near 1.5; compare, however, the discussion of Lyapunov exponents and dimensionality in Appendix 2.

A rather similar figure, composed of continous curves instead of randomly selected points on these curves, appears in Mandelbrot's comprehensive book *The Fractal Geometry of Nature;* it is actually these curves whose intersections with vertical lines are Cantor sets. The use of dots in Figure 64 is supposed to facilitate comparison with other figures constructed with dots, such as figures 12 and 16.

It was near the close of the seventies that "chaos" was rapidly becoming established as a standard term for phenomena exhibiting sensitive dependence. It was also at just about this time that new strange attractors were rapidly being encountered, and these attractors with

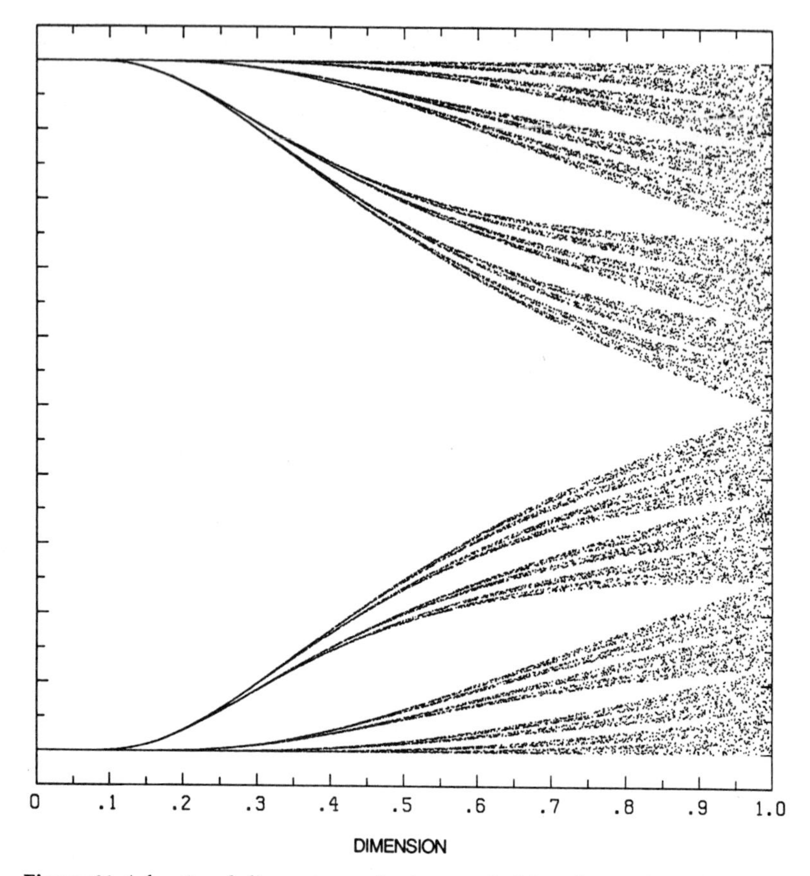

Figure 64. A fractional-dimension scale. Any vertical line drawn through the figure will intersect it in a Cantor set having the dimension, between 0.0 and 1.0, indicated at the base.

their fractal structure, rather than the absence of periodicity or the presence of sensitive dependence, were the features that some specialists were finding most appealing. Temporarily, at least, they were becoming the principal subject of a chaos theory. It was but a short step for "chaos" to extend its domain to fractals of all kinds, and even to more general shapes that had not become familiar objects of study before the advent of computers. In retrospect, it would be hard to maintain that the original meaning of "chaos" could more appropriately have been extended to one of these categories of shapes than another.

There is little question but that "chaos," like "strange attractor," is an appealing term—the kind that tends to establish itself. I have often speculated as to how well James Gleick's best-seller would have fared at the bookstores if it had borne a title like *Sensitive Dependence: Making a New Science*.

APPENDIX 1

The Butterfly Effect

THE FOLLOWING is the text of a talk that I presented in a session devoted to the Global Atmospheric Research Program, at the 139th meeting of the American Association for the Advancement of Science, in Washington, D.C., on December 29, 1972, as prepared for press release. It was never published, and it is presented here in its original form.

Predictability: Does the Flap of a Butterfly's Wings in Brazil Set off a Tornado in Texas?

Lest I appear frivolous in even posing the title question, let alone suggesting that it might have an affirmative answer, let me try to place it in proper perspective by offering two propositions.

1. If a single flap of a butterfly's wings can be instrumental in generating a tornado, so also can all the previous and subsequent flaps of its wings, as can the flaps of the wings of millions of other butterflies, not to mention the activities of innumerable more powerful creatures, including our own species.
2. If the flap of a butterfly's wings can be instrumental in generating a tornado, it can equally well be instrumental in preventing a tornado.

More generally, I am proposing that over the years minuscule disturbances neither increase nor decrease the frequency of occurrence of various weather events such as tornados; the most that they may do is to modify the sequence in which these events occur. The question which really interests us is whether they can do even this—whether, for example, two particular weather situations differing by as little as the immediate influence of a single butterfly will generally after sufficient time evolve into two situations differing by as much as the presence of a tornado. In more technical language, is the behavior of the atmosphere *unstable* with respect to perturbations of small amplitude?

The connection between this question and our ability to predict the weather is evident. Since we do not know exactly how many butterflies there are, nor where they are all located, let alone which ones are flapping their wings at any instant, we cannot, if the answer to our question is affirmative, accurately predict the occurrence of tornados at a sufficiently distant future time. More significantly, our general failure to detect systems even as large as thunderstorms when they slip between weather stations may impair our ability to predict the general weather pattern even in the near future.

How can we determine whether the atmosphere is unstable? The atmosphere is not a controlled laboratory experiment; if we disturb it and then observe what happens, we shall never know what would have happened if we had not disturbed it. Any claim that we can learn what would have happened by referring to the weather forecast would imply that the question whose answer we seek has already been answered in the negative.

The bulk of our conclusions are based upon computer simulation of the atmosphere. The equations to be solved represent our best attempts to approximate the equations actually governing the atmosphere by equations which are compatible with present computer capabilities. Generally two numerical solutions are compared. One of these is taken to simulate the actual weather, while the other simulates the weather which would have evolved from slightly different initial conditions, i.e., the weather which would have been predicted with a perfect forecasting technique but imperfect observations. The difference between the solutions therefore simulates the error in forecasting. New simulations are continually being performed as more powerful computers and improved knowledge of atmospheric dynamics become available.

Although we cannot claim to have proven that the atmosphere is unstable, the evidence that it is so is overwhelming. The most significant results are the following.

1. Small errors in the coarser structure of the weather pattern—those features which are readily resolved by conventional observing networks—tend to double in about three days. As the errors become larger the growth rate subsides. This limitation alone would allow us

to extend the range of acceptable prediction by three days every time we cut the observation error in half, and would offer the hope of eventually making good forecasts several weeks in advance.

2. Small errors in the finer structure—e.g., the positions of individual clouds—tend to grow much more rapidly, doubling in hours or less. This limitation alone would not seriously reduce our hopes for extended-range forecasting, since ordinarily we do not forecast the finer structure at all.
3. Errors in the finer structure, having attained appreciable size, tend to induce errors in the coarser structure. This result, which is less firmly established than the previous ones, implies that after a day or so there will be appreciable errors in the coarser structure, which will thereafter grow just as if they had been present initially. Cutting the observation error in the finer structure in half—a formidable task—would extend the range of acceptable prediction of even the coarser structure only by hours or less. The hopes for predicting two weeks or more in advance are thus greatly diminished.
4. Certain special quantities such as weekly average temperatures and weekly total rainfall may be predictable at a range at which entire weather patterns are not.

Regardless of what any theoretical study may imply, conclusive proof that good day-to-day forecasts can be made at a range of two weeks or more would be afforded by any valid demonstration that any particular forecasting scheme generally yields good results at that range. To the best of our knowledge, no such demonstration has ever been offered. Of course, even pure guesses will be correct a certain percentage of the time.

Returning now to the question as originally posed, we notice some additional points not yet considered. First of all, the influence of a single butterfly is not only a fine detail—it is confined to a small volume. Some of the numerical methods which seem to be well adapted for examining the intensification of errors are not suitable for studying the dispersion of errors from restricted to unrestricted regions. One hypothesis, unconfirmed, is that the influence of a butterfly's wings will spread in turbulent air, but not in calm air.

A second point is that Brazil and Texas lie in opposite hemispheres. The dynamical properties of the tropical atmosphere differ considerably from those of the atmosphere in temperate and polar latitudes. It is almost as if the tropical atmosphere were a different fluid. It seems entirely possible that an error might be able to spread many thousands of miles within the temperate latitudes of either hemisphere, while yet being unable to cross the equator.

We must therefore leave our original question unanswered for a few more years, even while affirming our faith in the instability of the atmosphere. Meanwhile, today's errors in weather forecasting cannot be blamed entirely nor even primarily upon the finer structure of weather patterns. They arise mainly from our failure to observe even the coarser structure with near completeness, our somewhat incomplete knowledge of the governing physical principles, and the inevitable approximations which must be introduced in formulating these principles as procedures which the human brain or the computer can carry out. These shortcomings cannot be entirely eliminated, but they can be greatly reduced by an expanded observing system and intensive research. It is to the ultimate purpose of making not exact forecasts but the best forecasts which the atmosphere is willing to have us make that the Global Atmospheric Research Program is dedicated.

APPENDIX 2

Mathematical Excursions

Numerical Integration

MOST OF THE FIGURES in this volume have been produced by a computer-graphics program, which simply plots sets of points and sometimes connects consecutive points with straight-line segments. The smooth curves that frequently appear have been formed by joining points that are very close together, or in some instances by plotting the points so closely that no connectors are needed. Except in schematic drawings, like the oblique view of the ski slope, the points to be plotted have been determined from appropriate systems of equations.

For mappings, the difference equations directly express future states in terms of present ones, and obtaining chronological sequences of points poses no problems. For flows, the differential equations must first be solved. General solutions of equations whose particular solutions are chaotic cannot ordinarily be found, and approximations to the latter are usually determined by numerical methods.

There are numerous procedures for numerical integration, but the "classical" fourth-order Runge-Kutta scheme—one of a family of schemes first devised at the end of the nineteenth century by the German mathematician Carl Runge, and brought to its present form a few years later by another German mathematician, Wilhelm Kutta—is especially popular, and when properly used can give excellent results. Except for the pinball system, all of the differential equations used as illustrative examples in this volume were integrated by this scheme.

To solve a typical system, say

$$dX/dt = F(X,Y),$$

$$dY/dt = G(X,Y),$$

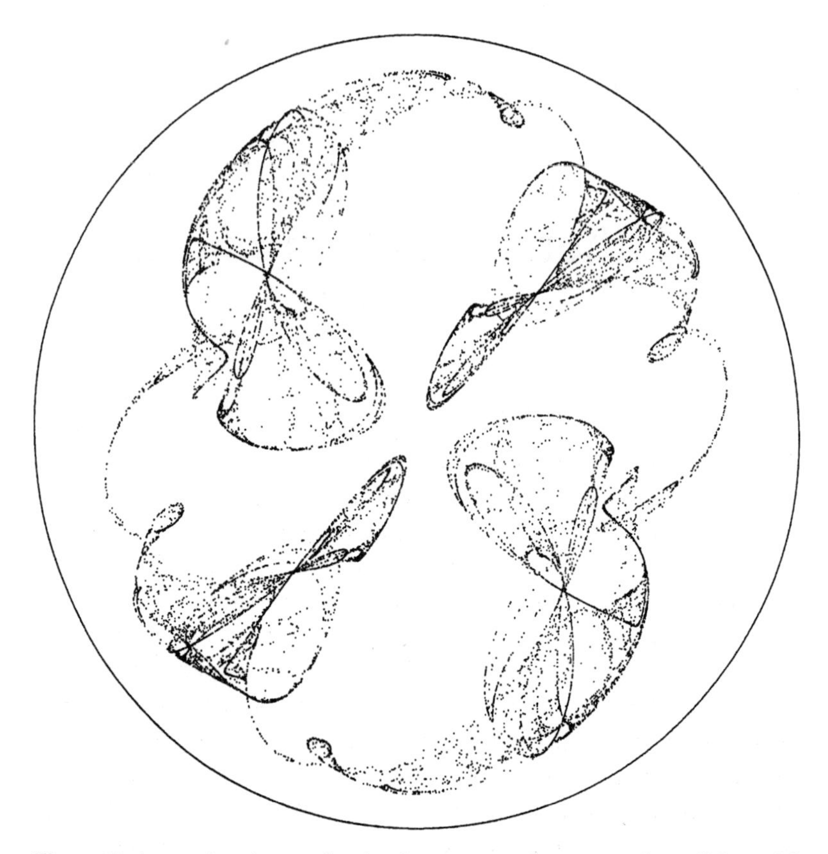

Figure 65. Approximations to the circular attractor of a system of two differential equations obtained when the equations are solved by the fourth-order Runge-Kutta scheme. The outer circle is the attractor with $\Delta t = 0.5$, and the inner structure is the attractor with $\Delta t = 1.65$.

choose a time increment Δt, and then, to find $X(t + \Delta t)$ and $Y(t + \Delta t)$ when $X(t)$ and $Y(t)$ are known at some time t, let

$$X_0 = X(t),$$
$$X_1 = X_0 + F(X_0, Y_0)\Delta t/2,$$
$$X_2 = X_0 + F(X_1, Y_1)\Delta t/2,$$
$$X_3 = X_0 + F(X_2, Y_2)\Delta t,$$
$$X_4 = X_0 - F(X_3, Y_3)\Delta t/2,$$

and finally,

$$X(t + \Delta t) = (X_1 + 2X_2 + X_3 - X_4)/3.$$

Analogous expressions hold for Y; note that both X_{i-1} and Y_{i-1} must be found before either X_i or Y_i can be evaluated. The procedure may be iterated as many times as desired, with the old value of $t + \Delta t$ used as the new value of t in each iteration.

The method gives highly accurate results when Δt is small enough, but bizarre things can happen if it is chosen too large. For example, the attractor of the system defined by the differential equations

$$dX/dt = X - Y - X^3$$

$$dY/dt = X - X^2Y$$

is a circle. With $\Delta t = 0.5$, the computed attractor, forming the periphery of Figure 65, is hard to distinguish from the correct circle, but, with $\Delta t = 1.65$, the circle gives way to the enclosed strange attractor. Incidentally, the strange attractor in Figure 53 was produced by applying a second-order Runge-Kutta scheme to the same set of equations; in this simpler scheme, X_0 and X_1 are defined as before, and

$$X(t + \Delta t) = X_0 + F(X_1, Y_1)\Delta t.$$

You may enjoy experimenting with different values of Δt.

Often we wish to interpolate between iterations; for example, to construct a Poincaré cross section we may need to know the value of Y at the time when $X = 0$. For $0 < c < 1$, a suitable formula is the fourth-degree polynomial approximation

$$\begin{aligned} X(t + c\Delta t) = X_0 &+ 2(X_1 - X_0)c + 2(X_2 - X_1)c^2 \\ &+ 2(X_3 - 2X_2 + X_0)c^3/3 - (X_4 + X_3 - X_1 - X_0)c^4/3, \end{aligned}$$

again with an analogous expression for Y. If we have observed that X has a zero crossing between t and $t + \Delta t$, we can set the expression for $X(t + c\Delta t)$ equal to zero, solve the resulting fourth-degree equation for c, and then evaluate $Y(t + c\Delta t)$. The equation for c is easily solved by Newton's method; let

$$c_0 = X(t)/(X(t) - X(t + \Delta t)),$$

and then, for $n = 0,1,\ldots,$ let

$$c_{n+1} = c_n - X(t + c_n \Delta t)/D(c_n),$$

where

$$D(c) = 2(X_1 - X_0) + 4(X_2 - X_1)c$$
$$+ 2(X_3 - 2X_2 + X_0)c^2 - 4(X_4 + X_3 - X_1 - X_0)c^3/3$$

is the derivative of the expression for $X(t + c\Delta t)$. When c_{n+1} becomes very close to c_n, stop, and let $c = c_{n+1}$. For some purposes c_0 may be a good enough approximation to c.

The Butterfly

The equations that produced the butterfly-shaped figure in Chapter 1, and that I used as the illustrative example of chaotic behavior in the paper "Deterministic Nonperiodic Flow," are

$$dx/dt = -\sigma x + \sigma y$$
$$dy/dt = -xz + rx - y,$$
$$dz/dt = xy - bz.$$

The three constants b, σ, and r determine the behavior of the system. The equations have been studied intensively, and Colin Sparrow of Cambridge University has even written an entire book about them.

The structure that has often been called the attractor is actually an extensive segment of a particular solution contained in the attractor. To obtain a butterfly like the one in the first chapter, let $b = 8/3$, $\sigma = 10$, and $r = 28$. Choose a suitable time step Δt, suitable initial values of x,y, and z, and solve the equations by some numerical procedure, such as the fourth-order Runge-Kutta scheme. Stop after a few thousand steps and plot the values of z against the corresponding values of x, omitting the first few points if they appear to represent transient conditions.

The butterfly has often been drawn with the successive points connected by line segments, so that it appears as a long continuous curve. The same effect can be obtained by choosing a very small time step. I originally encountered the new species of butterfly when I intended to let $\Delta t = 0.002$, but typed the decimal point in the wrong place. Mutations like this can sometimes produce superior creatures, but more often they produce garbage, and on occasions they can be disastrous.

Rössler's equations,

$$dx/dt = -y - z,$$
$$dy/dt = x + \alpha y,$$
$$dz/dt = \alpha + xz - \mu z$$

have only one instead of two nonlinear terms. The values used by Rössler, $\alpha = 0.2$ and $\mu = 5.7$, will produce chaos.

The Ski Slope

The equations for the motion of a board or a sled on a ski slope are simply expressions of Newton's law, and they equate acceleration to force per unit mass. If X, Y, Z are distances and U, V, W are velocity components in the southward, eastward, and upward directions, respectively, and if $H(X, Y)$ is the height of the slope above some horizontal reference plane, the equations are

$$dX/dt = U,$$
$$dY/dt = V,$$
$$dZ/dt = W,$$
$$dU/dt = -FH_X - cU,$$
$$dV/dt = -FH_Y - cV,$$
$$dW/dt = -g + F - cW,$$

where g is the acceleration of gravity, F is the vertical component of the force of the slope against the board or the sled, c is a coefficient of friction, and subscripts denote partial differentiation.

Since

$$Z = H(X, Y),$$

on the slope, it follows that

$$W = H_X U + H_Y V,$$

and

$$dW/dt = -H_X(FH_X + cU) - H_Y(FH_Y + cV) + (H_{XX}U^2 + 2H_{XY}UV + H_{YY}V^2).$$

Eliminating W and dW/dt, we find that

$$F = (g + H_{XX}U^2 + 2H_{XY}UV + H_{YY}V^2)/(1 + H_X^2 + H_Y^2),$$

and, with this value of F, the equations for X, Y, U, and V describe the motion.

For the board, c has been a prechosen constant, although, as noted before, it could have been made proportional to $F\sqrt{(1 + H_X^2 + H_Y^2)}$, the force of the slope against the board. For the sled, U has been a prechosen constant, and c has been chosen to make dU/dt vanish; thus

$$c = -FH_X/U.$$

In all of the examples in the text,

$$H = -aX - b\cos(pX)\cos(qY),$$

so that, in addition to g, and c or U, values of a, b, p, and q must be chosen before computations can begin. In each example, $2\pi/p = 10.0$ meters, $2\pi/q = 4.0$ meters, and $a = 0.25$ except in the Hamiltonian system, where $a = 0$. Generally $b = 0.5$ meters, except for specified cases where $b = 0.25$ meters, and in the discussion of bifurcation, where b ranges from 0.0 to 0.6 meters. Note that h, the height of a mogul above a neighboring pit, is simply $2b$. For the board, $c^{-1} = 2$ seconds, while for the sled, $U = 3.5$ meters per second.

For the sled, and for the board when it moves continually down the slope—this excludes the conservative system and the cases in which the board can become trapped in a pit—X may be used in place of t as the independent variable. The expressions for the time derivatives of Y and V, and U in the case of the board, are simply divided by U. We are left with a system of two equations for the sled or three for the board, with periodic dependence on the independent variable in either case, since $\cos(pX)$ and $\sin(pX)$ appear in the equations.

Volume-Preserving Chaos

When a Runge-Kutta scheme or some other scheme is used to solve a system of differential equations, it often produces a small but persistent injection or removal of energy, in addition to any increases or decreases that are actually demanded by the equations. If the system is dissipative, this effect does little more than slightly alter the intended rate of dissipation, which may have been chosen somewhat arbitrarily in any case. If on the other hand the system is nondissipative, the procedure may convert it into a dissipative system, with quite different long-term properties. It is therefore especially important in solving such equations to choose a very small time step, to cut down on any spurious behavior. This inevitably adds to the computation time.

If you are interested in exploring in more detail the scores of periodic islands that can dot a chaotic sea, you are strongly advised to use a system of difference equations, for which no special integration scheme is needed. One of the most popular of these systems is the so-called standard map, proposed by Boris Chirikov as a paradigm for Hamiltonian systems, and given in polar coordinates by

$$r_{n+1} = r_n + a \sin \theta_n,$$
$$\theta_{n+1} = \theta_n + r_{n+1}.$$

The simplest system of which I am aware, however, is the third-order difference equation

$$x_{n+1} = x_n x_{n-1} - x_{n-2},$$

where, given three successive terms, a single multiplication and a single subtraction will yield the next term. The equation preserves the quantity

$$Q = x_n^2 + x_{n-1}^2 + x_{n-2}^2 - x_n x_{n-1} x_{n-2};$$

thus, even though it contains no explicit constants, it effectively defines a family of dynamical systems—one for each value of Q. Surfaces of constant Q look much like spheres when Q is near zero, but become noticeably distorted when Q is larger, and open up and extend to infinity when $Q > 4$.

To observe the structure of a system, choose a value of Q; an intermediate value, perhaps between 1.0 and 2.0, should work well. Next choose values of x_1 and x_2, which should not be too large, and, treating the expression for Q with $n = 3$ as a quadratic equation in x_3, solve it; either of the two roots can be used as x_3, and the other one will be x_0. The points (x_{n-1}, x_n) will occupy the projection of the Q-surface on a plane, and, to avoid plotting the near side and the far side on top of each other, you can plot x_n against x_{n-1} only when $x_{n+1} > x_{n-2}$. You should be able to collect new points a thousand times as fast as with the volume-preserving slope model. With high resolution you may be able to discover chains of loops and chains of still smaller loops surrounding each of these loops, and perhaps something unexpected.

Hill's Reduced Problem

In Hill's reduced form of the three-body problem, all three bodies move in the same plane. One of them, the "satellite," has a negligibly small mass, and the other two, the "planets," travel about the center of the combined mass in circular paths.

If m_1 and m_2 are the masses of the planets and (x_1, y_1) and(x_2, y_2) are their positions in a Cartesian coordinate system, and if $x_{12} = x_2 - x_1$ and $y_{12} = y_2 - y_1$, and $r_{12}^2 = x_{12}^2 + y_{12}^2$, the equations governing their motion are

$$m_1 d^2x_1/dt^2 = cm_1 m_2 x_{12}/r_{12}^2,$$
$$m_2 d^2x_2/dt^2 = -cm_1 m_2 x_{12}/r_{12}^2,$$

with analogous equations for y_1 and y_2. Since the planets move in circular orbits, r_{12} is a constant, and the units of mass, distance, and time may be chosen so that $m_1 + m_2 = 1$, $r_{12} = 1$, and $c = 1$. Thus

$$d^2x_{12}/dt^2 = -x_{12}, \quad d^2y_{12}/dt^2 = -y_{12},$$

and, if the initial time is chosen so that $y_{12} = 0$ and $x_{12} > 0$,

$$x_{12} = \cos t, \quad y_{12} = \sin t,$$

whence

$$x_1 = -m_2 \cos t, \quad y_1 = -m_2 \sin t,$$
$$x_2 = m_1 \cos t, \quad y_2 = m_1 \sin t.$$

If (x, y) is the position of the satellite and if $r_1^2 = (x - x_1)^2 + (y - y_1^2)$ and $r_2^2 = (x - x_2)^2 + (y - y_2)^2$, the equations governing the motion are

$$d^2x/dt^2 = -m_1(x - x_1)/r_1^3 - m_2(x - x_2)/r_2^3,$$

again with an analogous equation for y. If new coordinates

$$X = x \cos t + y \sin t,$$
$$Y = -x \sin t + y \cos t$$

are introduced, so that the satellite is viewed in a coordinate system that rotates with the revolution of the planets, with the X-axis always passing through the planets, the equations may be written

$$dX/dt = U$$
$$dY/dt = V$$
$$dU/dt = X + 2V - m_1 X/r_1^3 - m_2 X/r_2^3 - m_1 m_2/r_1^3 + m_1 m_2/r_2^3,$$
$$dV/dt = Y - 2U - m_1 Y/r_1^3 - m_2 Y/r_2^3$$

with $r_1^2 = (X + m_2)^2 + Y^2$ and $r_2^2 = (X - m_1)^2 + Y^2$, so that, although the equations have not been shortened by the rotation, t no longer appears explicitly on the right-hand sides.

The equations preserve the value of the so-called Jacobi integral

$$J = (U^2 + V^2 - m_1/r_1 - m_2/r_2)/2,$$

so that effectively we have a family of three-variable dynamical systems, one for each value of J, instead of a single four-variable system. In this respect the system is like the conservative ski-slope model.

In order to preserve J when r_1 or r_2 becomes very small, U or V must become large, and a computation procedure like the Runge-Kutta scheme that will work well most of the time may suddenly fail when the satellite draws too near to one of the planets. To guard against this possibility it is advisable to let the length of the time step vary from step to step, making it very small when either r_1 or r_2 is very small.

A negative initial value of J will assure us that the satellite cannot escape to infinity. With some values of m_1, m_2, and J the satellite may be trapped in the vicinity of one planet or the other, while with others, including the ones used to produce Figures 29 and 30, it can shuttle between the planets.

A Poincaré section, say one where $Y = 0$ and $V > 0$, will produce a two-dimensional mapping. With fixed values of m_1, m_2, and J and various sets of initial conditions, you should be able to produce a diagram like Figure 20, with islands in a chaotic sea.

The Logistic Equation

The logistic equation is most commonly written

$$X_{n+1} = AX_n(1 - X_n);$$

the choice of symbols varies. For $0 < A \leq 4$, it defines a family of noninvertible mappings of the interval from 0 to 1 into itself. It is probably the simplest attainable equation for studying period-doubling bifurcations, with the main sequence beginning with a bifurcation from period 1 to period 2 when $A = 3$, and from period 2 to period 4 when $A = 1 + \sqrt{6} = 3.45$. It has served as the principal illustrative example in Robert Devaney's textbook *An Introduction to Chaotic Dynamical Systems.*

The values of A for which the behavior is periodic form an infinite number of finite intervals, while the values for which it is chaotic lie between 3.57 and 4.0 and form a Cantor set whose dimension appears to be 1.0. The Cantor set resembles the one that would be formed by taking an interval of unit length, removing the middle fourth, then removing the middle ninth of the two resulting pieces, then the middle sixteenth of the four pieces, etc. Here the sum of the lengths of the removed pieces is only ½.

If we substitute $c = A/2 - A^2/4$ and $z = A(1 - 2X)/2$, the equation becomes

$$z_{n+1} = z_n^2 + c,$$

another convenient form. The first two period-doubling bifurcations occur at $c = -3/4$ and $-5/4$. The values c_i where bifurcations from period 2^{i-1} to period 2^i occur converge to $c_\infty = -1.4012$, while the ratios $(c_\infty - c_{i-1})/(c_\infty - c_i)$ converge to 4.6692, a number discovered by Mitchell Feigenbaum to be characteristic of bifurcations occurring in a wide class of mappings and flows. Solutions of period 3 first appear at a saddle-node bifurcation at $c = -7/4$. Note that the equation is equivalent to the second-order conservative system

$$z_{n+1} = z_n + z_n^2 - w_n^2,$$

$$w_{n+1} = z_n,$$

where no constant enters explicitly but where $z - w^2$ retains its initial value.

When c and z are taken to be complex, the first-order equation becomes

$$x_{n+1} = x_n^2 + y_n^2 + a,$$
$$y_{n+1} = 2x_n y_n + b,$$

where $c = a + ib$ and $z = x + iy$. These equations generate the Mandelbrot set. The point (a, b) is in the set if the sequence (x_n, y_n) starting at $(0, 0)$ does not approach infinity. The familiar brilliantly colored pictures of the set are actually pictures of the region just outside the set. Different colors are assigned to different numbers of steps required for x or y to become very large, say $x^2 + y^2 = 10^6$. Any other large number may be used, provided that it will fit into the computer, since once x and y have become large they increase very rapidly. The choice of colors requires an artist's eye rather than a mathematician's.

Lyapunov Exponents and Dimensionality

Take an n-variable dynamical system. Choose a point, and take a small n-dimensional sphere centered at the point. As time increases, the sphere will be deformed into an approximate ellipsoid; compare Figure 15, where $n = 2$. In the limit as the *initial* diameter of the sphere approaches zero, the time during which the image will remain indistinguishable from an ellipsoid will approach infinity.

The long-term average factors by which the lengths of the axes of the ellipsoid are multiplied during one iteration, for a mapping, or one time unit, for a flow, are called *Lyapunov numbers,* and their logarithms are called *Lyapunov exponents.* That is, if in the long run an axis increases or decreases as rapidly as $\exp(\lambda t)$, the corresponding Lyapunov exponent is λ. Ordinarily the exponents are numbered in decreasing order, with $\lambda_1 \geq \ldots \geq \lambda n$.

If the exponents are to be characteristics of the system, certain choices of the original point must be avoided. These include unstable fixed points and points on unstable periodic solutions. If there are several attractors, there may be separate sets of exponents for the separate basins of attraction. If the system is Hamiltonian, there will be different exponents for the chaotic sea and the periodic solutions.

The sum of the exponents indicates the rate at which the volume of the ellipsoid will increase or decrease, so that it will be zero for Hamiltonian systems and negative for dissipative systems. If the attractor of a dissipative system is a fixed point, all of the exponents will generally be negative. If it is a simple m-dimensional manifold—a curve or a surface if $m = 1$ or 2—the first m exponents will be zero and the remaining ones will be negative. Chaos will occur, whether or not the system is dissipative, if $\lambda_1 > 0$.

Consider an n-dimensional box that encloses the attractor. The sum $\lambda_1 + \ldots + \lambda_k$ of the first k exponents indicates the rate at which the volume—length or area $k = 1$ if or 2—of the projection of an infinitesimal ellipsoid on a k-dimensional face of the box will increase or decrease. Thus, if $\lambda_1 > 0$, the projection of a small but finite ellipsoid on one edge of the box will continue to grow until such time as several points on the now highly distorted ellipsoid project onto the same point, i.e., until the projection folds over on itself. If in addition $\lambda_1 + \lambda_2 < 0$, the projection on a two-dimensional face will shrink. The attractor may then be expected to consist of a complex of curves, with no surfaces present.

If $\lambda_1 + \lambda_2 > 0$, whether or not $\lambda_2 > 0$, and if $\lambda_1 + \lambda_2 + \lambda_3 < 0$, the area of a projection on a two-dimensional face will continue to grow until folding occurs, while the volume of the projection on a three-dimensional face will continue to shrink, and the attractor may be expected to be composed of a complex of surfaces. More generally, if $\lambda_1 + \ldots + \lambda_k > 0$ but $\lambda_1 + \ldots + \lambda_{k+1} < 0$, the attractor should consist of a complex of k-dimensional manifolds.

A formula giving the fractional dimension of an attractor in terms of the associated Lyapunov exponents has been proposed by James Kaplan and James Yorke. Again if $\lambda_1 + \ldots + \lambda_k \geq 0$, and $\lambda_1 + \ldots + \lambda_{k+1} < 0$, implying that $\lambda_{k+1} < 0$, the dimension is

$$d = k + (\lambda_1 + \ldots + \lambda_k) / |\lambda_{k+1}|.$$

For some simple systems d can be shown to equal the capacity. For more involved systems where equality cannot be rigorously established, d may be accepted as an alternative definition of dimension.

For the Poincaré mapping of the sled model, if the unit of time is the time required to descend five meters, $\lambda_1 = 0.72$, so that the longest axis of the ellipsoid will double in just under five meters' descent, and

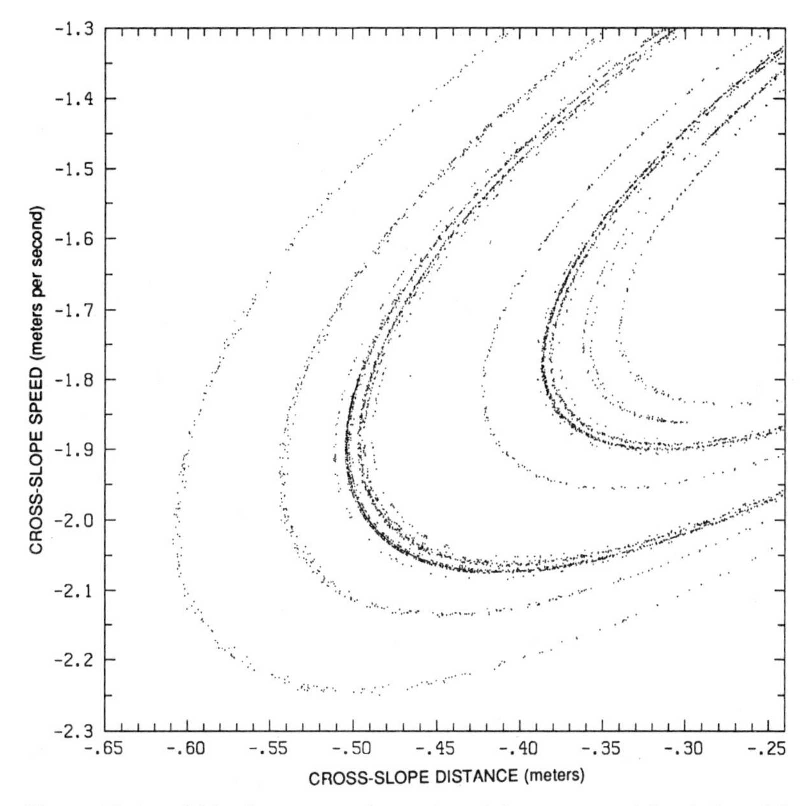

Figure 66. A tenfold enlargement of a section of the attractor of the sled model, shown in Figure 12.

$\lambda_2 = -1.53$, making $d = 1.47$. For the full flow the exponents are 0.72, 0, and –1.53. As is generally the case with flows for which the attractor is not a fixed point, one exponent is zero.

If all small circles behaved as the one in Figure 15 does initially, the Lyapunov numbers would be 4.8 and 0.047, so that the exponents would be 1.57 and –3.06. The chosen circle is therefore deformed about twice as rapidly as a typical circle, but it still provides a reasonable estimate, 1.51, of the dimension.

For the Poincaré mapping of the board model, $\lambda_1 = 0.67$, so that again an axis doubles its length in about one time unit, while $\lambda_2 = -0.70$ and $\lambda_3 = -1.36$, making $d = 1.96$. This value, almost high enough for the attractor to be composed of surfaces instead of curves, may appear sur-

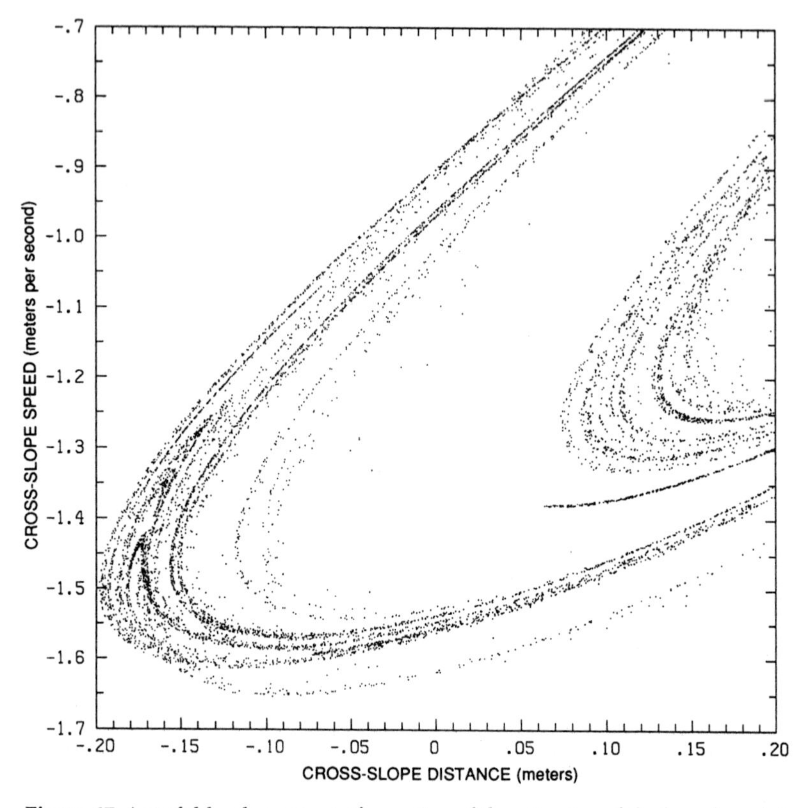

Figure 67. A tenfold enlargement of a section of the attractor of the board model, shown in Figure 16.

prising in view of the similarity between Figures 12 and 16. Recall, then, that dimensionality is defined in terms of limiting behavior as distances in phase space approach zero. It is thus a measure of fine structure rather than coarse structure. Often the attractor is assumed to be statistically self-similar, and the fine structure is not even investigated. In the present case, however, the two figures have similar coarse structures but noticeably different fine structures; in the sled attractor the points tend to be concentrated into curves, as in the center of the scale in Figure 64, while in the board attractor they have more of a tendency to spread out and fill areas, as on the right side of the scale. This becomes apparent in Figures 66 and 67, which show tenfold enlargements of similar sections of the two attractors.

Homoclinicity and the Horseshoe

Let us return to the sled sliding down the ski slope. Although its general behavior is chaotic, there are two simple periodic paths that it can follow, and it will nearly follow one for a while if we wait long enough. On these paths the sled moves exactly 2 meters eastward, or westward, for every 5 meters that it moves down the slope.

In a Poincaré mapping of V against y produced by observing the sled at 5-meter intervals, these paths show up as fixed points. Consider a small circle enclosing one of these points. Its successive images under the mapping will be closed curves also surrounding the fixed point, and the first few images will resemble ellipses. The long axes will continually grow, since the fixed point is unstable, while the enclosed areas, and hence the short axes, will continually diminish, since the system is dissipative. The infinitely long curve that the successive images will approximate more and more closely is called the *unstable manifold* of the fixed point, and will be denoted by $\boldsymbol{U}$. Any one of its points is mapped to a point that is farther from the fixed point when the distance is measured along $\boldsymbol{U}$, although its straight-line distance is often smaller. In this particular mapping, $\boldsymbol{U}$ is graphically indistinguishable from the attractor.

We may likewise construct the *stable manifold* $\boldsymbol{S}$ as the limiting form of successive inverse images of the circle. Points on $\boldsymbol{S}$ are mapped to points closer to the fixed point, and ultimately their forward images converge to the fixed point. Likewise the points on $\boldsymbol{U}$ emanate from the fixed point; that is, their inverse images converge to the fixed point. More general systems with more variables can have multidimensional unstable and stable manifolds.

Figure 68 shows segments of the two manifolds for one of the fixed points of the sled model, labeled O. The coordinate system is the perspective view of the inside of a cylinder, used in Figure 13. The noteworthy feature is that $\boldsymbol{U}$ and $\boldsymbol{S}$ intersect transversally, at the point labeled C.

Any sequence of successive images in either manifold is one of Poincaré's asymptotic solutions. Since the sequence of which C is a point lies in both manifolds, it is doubly asymptotic, and O is one of Poincaré's homoclinic points. It is evident, if we let D and E be the first two forward images of C, that $\boldsymbol{U}$ must intersect $\boldsymbol{S}$ again at D and then at E. Likewise, if B and A are the first two inverse images of C, $\boldsymbol{S}$ must intersect $\boldsymbol{U}$ again at B and then A. In fact, there is nothing more special about C than about A, B, D, or E.

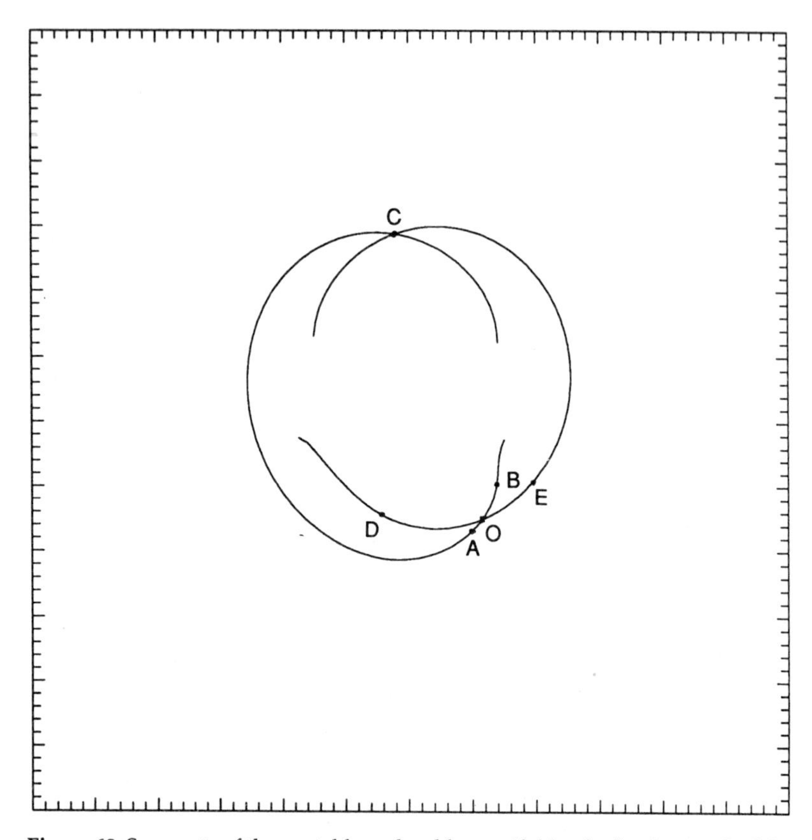

Figure 68. Segments of the unstable and stable manifolds of a fixed point O of the Poincaré mapping of the sled on the ski slope. The unstable manifold is shown passing through A and B while the stable manifold is shown passing through D and E, and the manifolds intersect transversally at C. The points B, C, D, and E are successive images of A. The coordinate system is that of Figures 13 and 37.

Although the distances between successive forward images of C, measured along S, continually decrease, the distances between these same points, measured along $\boldsymbol{U}$, continually increase. Likewise, the distances between successive inverse images, measured along S, continually increase. It follows that both $\boldsymbol{U}$ and S become continually more distorted as they are extended, and inevitably they collide in many points besides the images of C.

Note that in our example, and in fact in a great many systems, consecutive images of a point on either manifold, including points close to

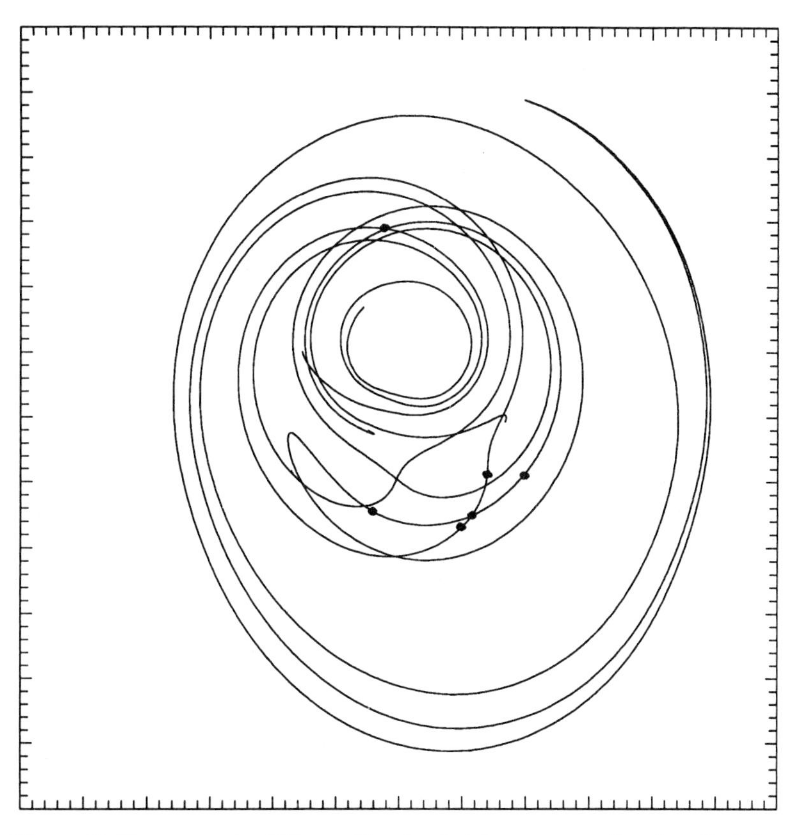

Figure 69. The segments of Figure 68, extended for one more iteration. Each intersection is of one manifold with the other. The dots indicate the points that have been labeled *A, B, C, D, E,* and *O* in Figure 68. The stable manifold is the one that extends farthest from the center and also closest to the center. The unstable manifold by itself, if extended for several more iterations, would be indistinguishable from Figure 37.

O, lie on opposite sides of *O*. Thus in the figure, since the left-hand extension of ***U*** from *O* passes through *C,* the right-hand extension must pass through *D,* while the left-hand extension passes through *E*. Likewise the right-hand extension of ***S*** passes through *C* and *A,* while the left-hand extension passes through *B*. This situation in no way affects the validity of Poincaré's arguments.

Extending ***S*** for many iterations is not computationally convenient in this example, since to do it we must integrate the differential equations backward in time. This effectively replaces friction by negative friction, and V can increase very rapidly, easily doubling its magnitude in one second. Since we are interested here only in the intersections of the manifolds, and since **U** never encounters values of V exceeding 5 meters per second, the high values of V on S are of no direct concern, and we can modify the equations to suppress the high values while leaving the lower values unchanged.

Figure 69 shows the two manifolds after the modification has been introduced and each manifold has been extended for one additional iteration. In places it may be hard to discern which manifold is which, but, since neither one intersects itself, all of the twenty-two intersections shown are of one manifold with the other.

To analyze the full significance of homoclinicity it is more convenient to turn to Smale's horseshoe. Figure 70 shows the structure in the form that Smale originally proposed, before it resembled a horseshoe closely enough to have acquired the name. To construct a particular example, begin with a square, whose sides lie on the lines $x = 0$ and $x = 1$ and whose bottom and top are on the lines $y = 0$ and $y = 1$. Divide the square into five sections by drawing the vertical lines $x = 0.3, 0.4, 0.6$, and 0.7. Compress the square vertically by a factor of ten, producing the elongated rectangle below the main figure. Then stretch the second and fourth sections, which by now have become small squares, horizontally by a factor of ten, stretch the middle section even more, compress the end sections, bend the stretched middle section, and finally fit the whole structure over the original square, as shown. What were the second and fourth vertical sections are now the second and fourth horizontal strips, and they lie between the lines $y = 0.3$ and 0.4, and $y = 0.6$ and 0.7. The equations of the mapping, for that part of the square that remains in the square, which is the only part that will concern us, are

$$x_{n+1} = 10x_n - 3, \quad y_{n+1} = (y_n + 3)/10 \quad \text{if} \quad 0.3 < x_n < 0.4,$$
$$x_{n+1} = 10x_n - 6, \quad y_{n+1} = (y_n + 6)/10 \quad \text{if} \quad 0.6 < x_n < 0.7.$$

The inverse mapping

$$x_{n-1} = (x_n + 3)/10, \quad y_{n-1} = 10y_n - 3 \quad \text{if} \quad 0.3 < y_n < 0.4,$$
$$x_{n-1} = (x_n + 6)/10, \quad y_{n-1} = 10y_n - 6 \quad \text{if} \quad 0.6 < y_n < 0.7$$

is the same as the forward one, with the roles of x and y interchanged.

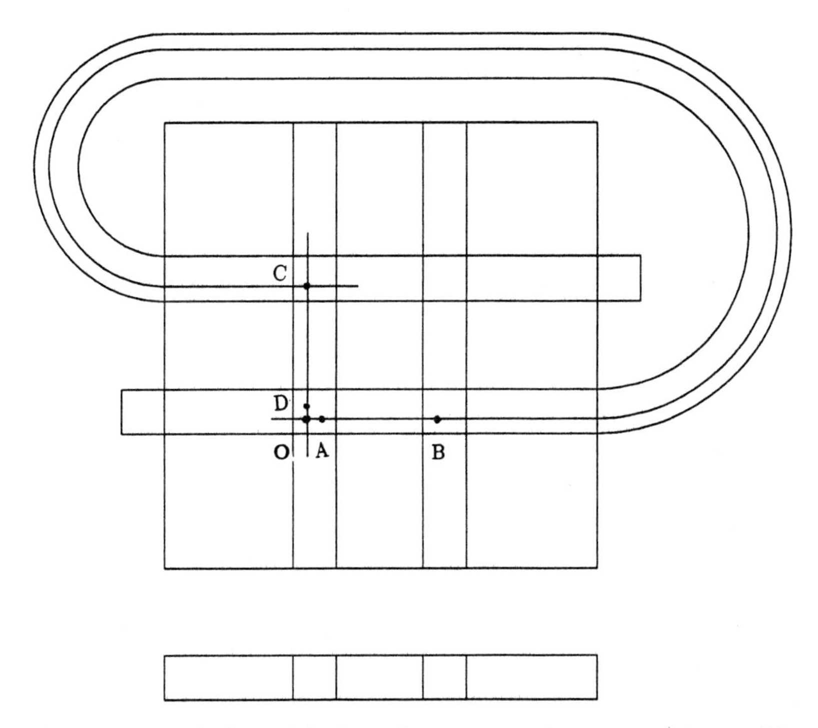

Figure 70. An early form of the horseshoe mapping. A segment of the unstable manifold of the fixed point *O*, extending horizontally through *A* and *B*, and a segment of the stable manifold, extending vertically through *D*, intersect transversally at *C*. The points *B*, *C*, and *D* are successive images of *A*.

The mapping possesses two fixed points, at ($\frac{1}{3}$, $\frac{1}{3}$) and ($\frac{2}{3}$, $\frac{2}{3}$). The unstable manifold of each point extends horizontally, becoming curved when it leaves the square, while the stable manifolds extend vertically. The figure shows segments of the two manifolds of point *O* at ($\frac{1}{3}$, $\frac{1}{3}$); these are seen to intersect transversally at point *C* at ($\frac{1}{3}$, $\frac{19}{30}$), so that *O* is homoclinic, whence the manifolds must intersect many more times—an infinite number, in fact.

Observe now that if we take any point originally in the second or fourth vertical section and write its coordinates in decimal form, its image is obtained by removing the leading digit, which must be 3 or 6, from x and attaching it to y; thus, for example, the image of (.3207, .549) is (.207, .3549). Likewise, an inverse image of an eligible point is obtained by transferring a digit from y to x. This sort of mapping, which simply transfers a leading digit, and shifts the positions of the others, is called a *Bernoulli shift*.

It follows that the points (x, y), all of whose forward and inverse images are in the square, are those where the decimal expressions for both x and y contain nothing but 3s and 6s. These points form an obvious two-dimensional Cantor set. Points on the stable manifold of O are those where there are only a finite number of 6s in the expression for x, so that, after a finite number of forward iterations, all of the digits of x are 3s, i.e., $x = 1/3$, and, after still more forward iterations, more and more leading digits of y are 3s, i.e., y approaches $1/3$. Points on the unstable manifold of O are those where y contains only a finite number of 6s. The points where both x and y contain a finite number of 6s are therefore the points where the manifolds intersect, and there are clearly an infinite number of these.

If the digits for y, preceded by the digits for x arranged in reverse order, form a periodic sequence, repeating say after n digits, the nth forward or inverse image of (x, y) will be (x, y) again, and (x, y) will lie on a periodic solution. Thus the number of periodic solutions is infinite. Randomly chosen infinite sequences of 3s and 6s will generally not be periodic, so that the number of nonperiodic solutions is also infinite. Clearly we are looking at chaos, at least in the limited sense. We can find solutions with special and perhaps peculiar properties simply by constructing unusual sequences of 3s and 6s. A solution with interesting behavior might be produced, for example, by letting the nth digit of x and also of y be 6 when n is a perfect square, and 3 otherwise.

It is because the mapping has been made linear in a part of the square that we can write explicit expressions for points on the unstable and stable manifolds and on periodic solutions. This would have still been so if we had used a simpler horseshoe, where the compressed, stretched, and bent square would simply exit from the original square on the right and then reenter on the right, although the mapping would involve some interchanging of 3s and 6s as well as shifts. The qualitative result is nevertheless topological; it would be unaltered by a continuous deformation of the mapping. If a horseshoe has the form of the one in Figure 41, or something considerably more distorted, it still follows that the number of intersections of the manifolds and the number of distinct periodic and also nonperiodic solutions will be infinite.

APPENDIX 3

A Brief Dynamical-Systems Glossary

THIS GLOSSARY presents some of the terminology of dynamical-systems theory that commonly appears in discussions of chaos. It is addressed primarily to the nonspecialist. It is intended to give the reader a feeling for what each concept is all about, and on this account the definitions appear in descriptive and sometimes colloquial form rather than in standard mathematical language. Some of the definitions lack mathematical precision, but the specialist should encounter little difficulty in converting them to rigorous statements.

As the only alternative to excessive repetition, most of the definitions contain terms that also appear as entries. Where one of these terms first enters a definition, it is shown in boldface type. It is anticipated that readers will use the glossary mainly to check the meanings of a few terms that they encounter in the course of reading the main text. The list is nevertheless intended to be self-contained, and the reader who wishes to start from scratch can, by rearranging or rereading the entries in a suitable nonalphabetic order, beginning with *system* and proceeding through *variable, constant,* and *state,* avoid encountering any terms not already defined. The reader who chooses to browse through the list in alphabetical order is warned that the concepts whose names begin with *A* tend to be the least simple.

The bulk of the definitions are presented in phase-space terminology. That is, *point* will generally be used to denote a state of a system, and *orbit* will denote a chronological sequence of states, or equivalently, if the system is defined by equations, a particular solution. Note that a fixed point is conventionally treated as a special case of an orbit.

The reader should be aware that many of the terms have additional meanings, including other technical meanings, when used in other contexts.

Almost-periodic orbit. An **orbit** that comes closer and closer to repeating its complete past history, after the passage of longer and longer fixed intervals of time. Compare **periodic orbit.**

Approach (a point or a **set**). To come near to and, ultimately, *remain* near to, within any prechosen degree of closeness. Colloquially, to draw nearer and nearer.

Approach infinity. To become larger and, ultimately, *remain* larger than any prechosen quantity.

asymptotic orbit. A **transient orbit** that **approaches** a **fixed point** or a **periodic** or **almost-periodic orbit.**

Attracting set. In a **dissipative system,** the set consisting of the **limit sets** of all **orbits,** together with all points on orbits that emanate from this set.

Attractor. In a **dissipative system,** a **limit set** that is not contained in any larger limit set, and from which no **orbits emanate.**

Axis. Any one of a particular set of mutually perpendicular lines passing through the **origin,** used as reference lines.

Basin of attraction. The **set** consisting of all **points** lying on **orbits** that **approach** a given **attractor.**

Bifurcation. In a **family** of **dynamical systems,** an abrupt change in the long-term behavior of a system, when the value of a **constant** is changed from below to above some critical value.

Butterfly effect. The phenomenon that a small alteration in the **state** of a **dynamical system** will cause subsequent states to differ greatly from the states that would have followed without the alteration; **sensitive dependence.**

Cantor set. 1. A **set** of points on a line or a curve such that, between any two points, there are other points of the set and also gaps of finite width.

2. A generalization of a Cantor set, as defined above, to more than one dimension.

Capacity. A particular measure of the **dimension** of a **set,** based upon the rate at which the number of cubes or spheres needed to cover the set increases, as the diameter of each cube or sphere decreases.

Chaos. 1. The property that characterizes a **dynamical system** in which most **orbits** exhibit **sensitive dependence;** full chaos.

2. Limited chaos; the property that characterizes a dynamical system in which some special orbits are **nonperiodic** but most are **periodic** or **almost periodic.**

Chaotic sea. The **set approached** by a chaotic orbit in a **Hamiltonian system.**

Compact system. A **dynamical system** in which every **orbit** possesses a **limit set.**

Completely random. See **random.**

Conservative system. A **dynamical system** in which some ostensibly variable quantity actually remains constant as time progresses.

Constant (of a **system**). A feature that does not vary as time progresses.

Coordinate (of a point). The distance from the **origin** to the closest point, on a particular **axis,** to the given point.

Deterministic system. A system in which later **states** evolve from earlier ones according to a fixed law.

Difference equation. An equation that expresses the value of a **variable** of a **system,** at a time following a given time, in terms of the values of all of the variables at the given time.

Differential equation. An equation that expresses the rate at which a **variable** of a **system** is changing at a given time, in terms of the values of all of the variables at that time.

Dimension. Any one of a number of measures of a **set** of points that agrees with the classical concept of dimension when the set is a point, curve, surface, or other manifold, but is also defined, often as a fraction, for more general sets.

Dissipative system. A **dynamical system** in which the image of any **set** of **points** of finite volume in **phase space** is a set of smaller volume.

Doubly asymptotic orbit. An **orbit** that is **asymptotic** to a **fixed point** or a **periodic orbit** and also **emanates from** a fixed point or a periodic orbit.

Dynamical system. A **deterministic system.** Also, liberally, a system with a slight amount of randomness, provided that the qualitative behavior would not be appreciably changed if the randomness were somehow removed.

Emanate from. To **approach,** if one travels in the reverse direction along an **orbit.**

Equilibrium. A **fixed point,** or, sometimes, a **periodic orbit.**

Family. A set of **dynamical systems** that are alike except for the values of one or more **constants.**

Fixed point. A **point** that is identical with its own **image.**

Flow. A **dynamical system** whose **variables** are defined for continuously varying values of time. A flow is often governed by a set of **differential equations.**

Fractal. A **set** of points whose **dimension** is not a whole number. Also, a set of similar structure whose dimension "happens to be" a whole number.

Full chaos. See **chaos.**

Hamiltonian system. A certain type of **conservative volume-preserving system.**

Homoclinic point. The **fixed point** from which a **homoclinic orbit emanates** and which it subsequently **approaches.**

Homoclinic orbit. An **orbit** that is **asymptotic** to a **fixed point** or a **periodic orbit** and **emanates from** the same point or orbit.

Horseshoe. A particular type of two-dimensional **mapping,** in which a square or some other area is mapped into a distorted area that intersects the original area in two disjoint pieces.

Image. The **set** of **points** that follows a given set by a specified number of iterations, one iteration unless otherwise stated, for a **mapping,** or by a specified amount of time, for a **flow.**

Initial conditions. The **state** of a **system,** at the beginning of any stretch of time that may be of interest to an investigator.

Invariant set. A set of **points** that is identical with its own **image.**

Inverse image. The set of **points** whose **image** consists of a given set.

Invertible system. A **dynamical system** in which each **point** has one and only one **inverse image.**

Limit set (of an **orbit**). A **set** that is **approached** by an orbit, and does not contain a smaller set approached by the orbit. Colloquially, the set consisting of every point that the orbit passes very close to, again and again.

Limited chaos. See **chaos.**

Linear system. A **system** in which alterations in an **initial state** will result in proportional alterations in any subsequent state.

Logistic equation. A particular quadratic **difference equation** in one variable.

Lyapunov exponents. The logarithms of the **Lyapunov numbers.**

Lyapunov numbers. The long-term average factors by which the lengths of the **axes** of an infinitesimal ellipsoid in **phase space** are multiplied, when the ellipsoid is replaced by its successive **images.**

Manifold. A point, curve, surface, or volume, or its generalization in multidimensional space.

Mapping. A **dynamical system** whose **variables** are defined only for discrete values of time. A mapping is often governed by a set of **difference equations.**

Model. A **system** designed to possess some of the properties of another, generally more complicated, system.

Noninvertible system. A **dynamical system** in which some **points** have several or no **inverse images;** a system that is not **invertible.**

Nonlinear system. A **system** in which alterations in an **initial state** need not produce proportional alterations in subsequent states; one that is not **linear.**

Nonperiodic orbit. An **orbit** where any sufficiently close repetition of a past **state** is of temporary duration; an orbit that is neither **periodic** nor **almost periodic.**

Orbit. The representation in **phase space** of a continuous or discrete chronological sequence of **states.**

Origin. A particular point in **phase space** or ordinary space, used as a reference point.

Parameter. A **constant** whose value can differ from one member of a **family** of **dynamical systems** to another.

Period. The number of iterations or the interval of time between successive repetitions in a **periodic orbit.**

Period-doubling bifurcation. 1. A **bifurcation** from a **system** in which typical **orbits** are **periodic** to one in which typical orbits are also periodic, but with a **period** twice as long.

2. An infinite sequence of period-doubling bifurcations, as defined above, culminating in **chaotic** behavior.

Periodic orbit. An **orbit** that exactly repeats its past behavior after the passage of a fixed interval of time.

Periodic system. A **system** in which all but a few exceptional **orbits** are **periodic** or **almost periodic,** or are **asymptotic** to periodic or almost-periodic orbits.

Periodic window. In a **family** of **dynamical systems,** a continuous set of values of a **parameter** for which the corresponding system is not chaotic, separating values for which the system is chaotic.

Phase space. A hypothetical space having as many dimensions as the number of **variables** needed to specify a state of a given **dynamical system.** The coordinates of a **point** in phase space are a set of simultaneous values of the variables.

Poincaré mapping. A mapping whose **phase space** is a **Poincaré section** of the phase space of a **flow,** and where successive images of a point are successive intersections of an **orbit** in the flow with the Poincaré section.

Poincaré section. A cross section of the **phase space** of a **flow** that is intersected by many or most **orbits.**

Point. The representation in **phase space** of a **state** of a **dynamical system.**

Random system. 1. A **system** in which the progression from earlier to later states is not completely determined by any law; a system that is not **deterministic.**

2. A system in which later states occur completely independently of earlier states; a completely random system.

Self-similar set. A **set** of which a portion, if magnified, becomes identical to the original set.

Sensitive dependence. The property characterizing an **orbit** if most other orbits that pass close to it at some **point** do not remain close to it as time advances.

Separatrix. A boundary separating two **basins of attraction.**

Set (of points). Any collection of points; often, a curve, surface, or some other structure, treated as an aggregation of points.

Stable equilibrium. A **fixed point** or a **periodic orbit** from which no **orbits emanate.**

Stable manifold. A **manifold** formed by the set of **orbits** that approach a given **fixed point** or **periodic orbit.**

State. The condition of a system at one instant; a set of simultaneous values of the **variables** of a system.

Statistically self-similar set. A **set** of which a portion, if magnified, has the same typical structure as the original set.

Strange attractor. An **attractor** with a **fractal** structure; one whose intersection with a suitable **manifold** is a **Cantor set.**

Surface of section. A **Poincaré section.**

System. Any entity that can undergo variations of some sort as time progresses.

Transient orbit. An **orbit** that has no **points** in common with its **limit set.**

Unstable equilibrium. A **fixed point** or a **periodic orbit** from which at least one **orbit emanates;** an **equilibrium** that is not **stable.**

Unstable manifold. A **manifold** consisting of all **points** on **orbits** that **emanate from** a given **fixed point** or **periodic orbit.**

Variable (of a **system**). A feature that can vary as time progresses.

Volume-preserving system. A **dynamical system** in which the image of any **set** of **points** in **phase space** is a set having the same volume.

Bibliography

THE VOLUME of literature devoted to chaos has grown to the point where it would be difficult for anyone to become familiar with all or even most of it. It continues to expand. Hao Bai-lin has supplemented his extensive collection of reprinted articles in *Chaos* and *Chaos II* with a selected bibliography of more than two thousand entries, while Zhang Shu-yu, a protégée of Hao's, has subsequently listed 7,460 items, including 303 books, in her *Bibliography on Chaos,* published in 1991.

In assembling the present list, I have rejected any thought of undertaking the arduous and perhaps impractical task of picking a small but representative sample, or of trying to identify the "important" items. The inevitable result has been that the selection, like the text, is somewhat slanted toward the works with which I am more familiar. The items finally chosen, presented as a single list, have been restricted to three categories. Some items in the first or second category fit into the third as well.

The first category consists of a few wholly or largely nontechnical works, which should provide many of you with some enjoyable recreational reading. These works extend from Henri Poincaré's famous set of essays, *Science and Method,* to the recent rather tersely titled books by John Casti, Ivar Ekeland, James Gleick, Benoit Mandelbrot, Heinz-Otto Peitgen and Peter Richter, Ivars Peterson, David Ruelle, and Ian Stewart, giving descriptive accounts of chaotic phenomena. They also include the historical accounts by the meteorologists George Platzman and Philip Thompson, and the brief autobiographical item by Stephen Smale.

Next are some textbooks and other book-length works of a more technical nature, whose primary aim is to expose the learner to the principles of dynamical systems and chaos, rather than to present any original findings. They include the volumes by Abraham and Shaw, Bergé et al., Birkhoff, Cvitanovic, Devaney, Glass and Mackey, Guckenheimer and Holmes, Gumowski and Mira, Holden, Lichtenberg and Lieberman, Moon, Nemytskii and Stepanov, Ruelle, Schuster, and Swinney and Gollub, as well as the extended exposition by Smale. Ruelle's *Chance and Chaos* falls into the first category, while his *Chaotic Evolution and Strange Attractors* belongs in the second.

Finally there is the largest category, consisting of the works that I have specifically referred to, or often merely alluded to, in the text of this volume. They form a rather heterogeneous set. The majority are scientific papers describing original research, but some are survey articles or books, and there are even a short story by Ray Bradbury and a novel by George R. Stewart.

The list follows.

Abraham, R. H., and C. D. Shaw. *Dynamics; the Geometry of Nature.* Rewood City, CA: Addison-Wesley, 1982.

Barnsley, M. F. "Making Chaotic Dynamical Systems to Order." In *Chaotic Dynamics and Fractals,* ed. M. F. Barnsley and S. G. Demko, pp. 53–68. Orlando: Academic Press, 1986.

Beddington, J. R., C. A. Free, and J. H. Lawton. "Dynamic Complexity in Predator-Prey Models Framed in Difference Equations." *Nature* 255 (1975): 58.

Bergé P., Y. Pomeau, and C. Vidal. *Order Within Chaos.* New York: Wiley, 1984.

Birkhoff, G. D. *Dynamical Systems.* Providence: American Mathemathical Society, 1926.

Birkhoff, G. D. "Sur Quelques Courbes Fermées Remarquables." *Bull. Soc. Math. France* 60 (1932): 1–26.

Bjerknes, V. "Das Problem der Wettervorhersage, betrachtet vom Standpunkte der Mechanik und der Physik." *Meteorol. Zeitschr.* 21 (1904): 1–7.

Bjerknes, V. "Application of Line Integral Theorems to the Hydrodynamics of Terrestrial and Cosmic Vortices." *Astrophys. Norvegica* 2, no. 6 (1937): 263–339.

Bradbury, R. D. "A Sound of Thunder." In *The Stories of Ray Bradbury,* pp. 231—241. New York: Alfred A. Knopf, 1980.

Cantor, G. "Grundlagen einer allgemeinen Mannichfältigkeitslehre." *Math. Annalen* 21 (1883): 545–591.

Cartwright, M. L., and J. E. Littlewood. "On Non-linear Differential Equations of the Second Order: I. The equation $\ddot{y}$-$k(1$-$y^2)\dot{y} + y = b\lambda\cos(\lambda t + \alpha)$, k large." *J. London Math. Soc.* 20 (1945): 180–189.

Casti, J. *Searching for Certainty.* New York: Morrow, 1990.

Charney, J. G., R. G. Fleagle, V. E. Lally, H. Riehl, and D. Q. Wark. "The Feasibility of a Global Observation and Analysis Experiment." *Bull. Amer. Meteorol. Soc.* 47 (1966): 200–220.

Chirikov, B. V., and F. M. Izrailev. "Degeneration of Turbulence in Simple Systems." *Physica D* 2 (1981): 30–37.

Cvitanovic, P. *Universality in Chaos: a Reprint Selection.* Bristol: Adam Hilger, 1984.

Devaney, R. L. *An Introduction to Chaotic Dynamical Systems.* Menlo Park, CA: Benjamin/Cummings, 1986.

Duffing, G. *Erzwangene Schwingungen bei veränderlicher Eigenfrequenz und ihre technische Bedeutung.* Braunschweig: Vieweg, 1918.

Ekeland, I. *Mathematics and the Unexpected.* Chicago: University of Chicago Press, 1988.

Faller, A. "A Demonstration of Fronts and Frontal Waves in Atmospheric Models." *J. Meteorol.* 13 (1956): 1–4.

Feigenbaum, M. "Quantitative Universality for a Class of Nonlinear Transformations." *J. Statistical Phys.* 19 (1978): 25–52.

Ford, J. "How Random is a Coin Toss." *Physics Today,* 36, no. 4 (1983): 40–47.

Franklin, P. "Almost Recurrent Periodic Motions." *Math. Zeitschr.* 30 (1929): 325–331.

Fultz, D., R. R. Long, G. V. Owens, W. Bohan, R. Kaylor, and J. Weil. "Studies of Thermal Convection in a Rotating Cylinder with Some Implications for Large-Scale Atmospheric Motion." *Meteorol. Monographs* (American Meteorological Society) 21, no. 4 (1959).

Glass, L., and M. C. Mackey. *From Clocks to Chaos.* Princeton: Princeton University Press, 1988.

Gleick, J. *Chaos: Making a New Science.* New York: Viking Penguin, 1987.

Guckenheimer, J. "A Strange, Strange Attractor." In *The Hopf Bifurcation and Its Applications,* ed. J. E. Marsden and M. McCracken, pp. 368–381. New York: Springer-Verlag, 1976.

Guckenheimer, J., and P. Holmes. *Nonlinear Oscillations, Dynamical Sys-*

tems, and Bifurcations of Vector Fields. New York: Springer-Verlag, 1983.

Gumowski, I., and C. Mira. *Recurrences and Discrete Dynamic Systems*. Lecture Notes in Mathematics, 809. New York: Springer-Verlag, 1980.

Haken, H. "Analogy between Higher Instabilities in Fluids and Lasers." *Phys. Letters*. 53A (1975): 77.

Hao, B.-L. *Chaos*. Singapore: World Scientific, 1984.

Hao, B.-L. *Chaos II*. Singapore: World Scientific, 1990.

Hide, R. "Some Experiments on Thermal Convection in a Rotating Liquid." *Quart. J. Roy. Meteorol. Soc.* 79 (1953): 161.

Hill, G. W. "Researches in the Lunar Theory." *Amer. J. Math.* 1 (1878): 5–26.

Holden, A. V., ed. *Chaos*. Princeton: Princeton University Press, 1986.

Kaplan, J. L., and J. A. Yorke. "Chaotic Behavior of Multidimensional Difference Equations." In *Functional Differential Equations and Approximation of Fixed Points*, ed. H.-O. Peitgen and H. O. Walther, pp. 204–227. Lecture Notes in Mathematics, 730. Berlin: Springer-Verlag, 1979.

Keefe, D. H., and B. Laden. "Correlation Dimension of Woodwind Multiphonic Tones." *J. Acoustical Soc. Amer.* 90 (1991): 1754–1765.

Kutta, W. "Beitrag zur naherungsweisen Integration totaler Differentialgleichungen." *Zeit. Math. Phys.* 46 (1901): 435–453.

Li, T. Y., and J. A. Yorke. "Period Three Implies Chaos." *Amer. Math. Monthly*. 82 (1975): 985–92.

Libchaber, A., and J. Maurer. "Local Probe in a Rayleigh-Bénard Experiment in Liquid Helium." *J. Physique—Lettres* 39 (1978): L-369–L-372.

Lichtenberg, A. J., and M. A. Lieberman. *Regular and Stochastic Motion*. New York: Springer-Verlag, 1983.

Lorenz, E. N. "The Statistical Prediction of Solutions of Dynamic Equations." *Proc. Internat. Sympos. Numer. Weather Pred.* Tokyo: Meteorological Society Japan (1962): 629–635, 647.

Lorenz, E. N. "Deterministic Nonperiodic Flow." *J. Atmos. Sci.* 20 (1963): 130–141.

Lorenz, E. N. "Atmospheric Predictability Experiments with a Large Numerical Model." *Tellus* 34 (1982): 505—513.

Lorenz, E. N. "Irregularity: a Fundamental Property of the Atmosphere." *Tellus* 36A (1984): 98—110.

Mandelbrot, B. B. "How Long Is the Coast of Britain? Statistical Self-similarity and Fractional Dimension." *Science* 156 (1967): 636—638.

Mandelbrot, B. B. *The Fractal Geometry of Nature*. San Francisco: W. H.

Freeman, 1982.

Markov, A. A. "Stabilität im Liapounoffschen Sinne und Festperiodizität," *Math. Zeitschr.* 36 (1933): 708–738.

May, R. M. *Stability and Complexity in Model Ecosystems.* Princeton: Princeton University Press, 1973.

May, R. M. "Simple Mathematical Models with Very Complicated Dynamics." *Nature* 261 (1976): 459–467.

McDonald, S. W., C. Grebogi, E. Ott, and J. A. Yorke. "Fractal Basin Boundaries." *Physica D* 17 (1985): 125–153.

Moon, F. C. *Chaotic Vibrations.* New York: Wiley, 1987.

Nemytskii, V. V., and V. V. Stepanov. *Qualitative Theory of Differential Equations.* Princeton: Princeton University Press, 1960.

Newton, I. *Philosophiae Naturalis Principia Mathematica.* London: S. Pepys, 1686.

Peitgen, H.-O., and P. H. Richter. *The Beauty of Fractals.* New York: Springer-Verlag, 1986.

Peterson, I. *The Mathematical Tourist.* New York: Freeman, 1988.

Phillips, N. "The General Circulation of the Atmosphere: a Numerical Experiment." *Quart. J. Roy. Meteorol. Soc.* 82 (1956): 123–164.

Platzman, G. W. "The ENIAC computations of 1950—Gateway to Numerical Weather Prediction." *Bull. Amer. Meteorol. Soc.* 60 (1979): 302–312.

Platzman, G. W. "Charney's Recollections." In *The Atmosphere—a Challenge,* ed. R. S. Lindzen, E. N. Lorenz, and G. W. Platzman, pp. 11–85. Boston: American Meteorological Society, 1990.

Poincaré H. *Les Méthodes Nouvelles de la Mécanique Céleste.* Paris: Gauthiers-Villar, 1893.

Poincaré H. *Science et Méthode.* Paris: Flammarion, 1912. English translation: *Science and Method.* Lancaster, PA: Science Press, 1913.

Prigogine, I., and I. Stengers. *Order Out of Chaos.* New York: Bantam, 1984.

Reed, R. "The Present Status of the 26-month Oscillation." *Bull Amer. Meteorol. Soc.* 46 (1965): 374–387.

Richardson, L. F. *Weather Prediction by Numerical Process.* Cambridge: Cambridge University Press, 1922.

Robbins, K. A. "A Moment Equation Description of Magnetic Reversals in the Earth." *Proc. Nat. Acad. Sci. U.S.A.* 73 (1976): 4297–4301.

Rössler O. E. "An Equation for Continuous Chaos." *Physics Letters A* 57

(1976): 397–398.

Ruelle, D. "Strange Attractors." *Math. Intelligencer,* 2, no. 126 (1980): 37–48.

Ruelle, D. *Chaotic Evolution and Strange Attractors.* Cambridge: Cambridge University Press, 1989.

Ruelle, D. *Chance and Chaos.* Princeton: Princeton University Press, 1991.

Ruelle, D., and F. Takens. "On the Nature of Turbulence." *Commun. Math. Phys.* 20 (1971): 167–192.

Runge, C. " Über die numerische Auflösing von Differentialgleichungen." *Math. Annalen* 46 (1895): 167–178.

Saltzman, B. "Finite Amplitude Free Convection as an Initial Value Problem—I." *J. Atmos. Sci.* 19 (1962): 329–341.

Schuster, H. G. *Deterministic Chaos.* Weinheim: Physik-Verlag, 1984.

Segel, L. A. "The Structure of Non-linear Cellular Solutions to the Boussinesq Equations." *J. Fluid Mech.* 21 (1965): 345–358.

Shaw, N. *Manual of Meteorology.* Cambridge: Cambridge University Press, 1926.

Smale, S. "Differentiable Dynamical Systems." *Bull. Amer. Math. Soc.* 73 (1967): 747–817.

Smale, S. "How I Got Started in Dynamical Systems." In *The Mathematics of Time,* by S. Smale, pp. 147–151. New York: Springer-Verlag, 1980.

Sparrow, C. *The Lorenz Equations: Bifurcations, Chaos, and Strange Attractors.* New York: Springer-Verlag, 1982.

Stewart, G. R. *Storm.* New York: Random House, 1941.

Stewart, I. *Does God Play Dice?* Oxford: Basil Blackwood, 1989.

Swinney, H. L. "Observations of Order and Chaos in Nonlinear Dynamics." *Physica D* 7 (1983): 3–15.

Swinney, H. L., and J. P. Gollub, eds. *Hydrodynamic Instabilities and the Transition to Turbulence.* Berlin: Springer-Verlag, 1987.

Thompson, P. D. "A History of Numerical Weather Prediction in the United States." *Bull. Amer. Meteor. Soc.* 64 (1983): 755–769.

Ueda, Y. "Strange Attractors and the Origin of Chaos." In *The Road to Chaos,* by Y. Ueda, pp. 185–216. Santa Cruz, CA: Aerial Press, 1992.

Ueda, Y. "Survey of Regular and Chaotic Phenomena in the Forced Duffing Oscillator." *Chaos, Solitons and Fractals* 1 (1991): 199–231.

van der Pol, B. "On 'Relaxation-Oscillations.'" *Phil. Magazine, 7th Ser.* 2 (1926): 978–992.

von Neumann, J. *Theory of Self-Reproducing Automata.* Edited by A. W.

Burks. Urbana: University of Illinois Press, 1966.

Welander, P. "On the Oscillatory Instability of a Differentially Heated Fluid Loop." *J. Fluid Mech.* 29 (1967): 17–30.

Wiener, N. "Nonlinear Prediction and Dynamics." *Proc. 3rd Berkeley Sympos. Math. Stat. and Prob.* (1956): 247–252.

Williams, R. F. "Lorenz Knots are Prime." *Ergodic Theor. and Dyn. Systems* 4 (1982): 147–163.

Wolfram, S. "Universality and Complexity in Cellular Automata." *Physica D* 10 (1984): 1–35.

Zhang, S.-Y. *Bibliography on Chaos.* Singapore: World Scientific, 1991.

Index

Page numbers are shown in boldface type when the reference is to a figure.